Starting a Duck Farm

A Collection of Articles on Stock Selection, Rearing, Economics and Other Aspects of Duck Farming

By

Various Authors

Copyright © 2013 Read Books Ltd.
This book is copyright and may not be
reproduced or copied in any way without
the express permission of the publisher in writing

British Library Cataloguing-in-Publication Data
A catalogue record for this book is available from the
British Library

Poultry Farming

Poultry farming is the raising of domesticated birds such as chickens, turkeys, ducks, and geese, for the purpose of farming meat or eggs for food. Poultry are farmed in great numbers with chickens being the most numerous. More than 50 billion chickens are raised annually as a source of food, for both their meat and their eggs. Chickens raised for eggs are usually called 'layers' while chickens raised for meat are often called 'broilers'. In total, the UK alone consumes over 29 million eggs per day

According to the Worldwatch Institute, 74% of the world's poultry meat, and 68% of eggs are produced in ways that are described as 'intensive'. One alternative to intensive poultry farming is free-range farming using much lower stocking densities. This type of farming allows chickens to roam freely for a period of the day, although they are usually confined in sheds at night to protect them from predators or kept indoors if the weather is particularly bad. In the UK, the Department for Environment, Food and Rural Affairs (Defra) states that a free-range chicken must have day-time access to open-air runs during at least half of its life. Thankfully, free-range farming of egg-laying hens is increasing its share of the market. Defra figures indicate that 45% of eggs produced in the UK throughout 2010 were free-range, 5% were produced in barn systems and 50% from cages. This compares with 41% being free-range in 2009.

Despite this increase, unfortunately most birds are still reared and bred in 'intensive' conditions. Commercial hens usually begin laying eggs at 16–20 weeks of age, although production gradually declines soon after from approximately 25 weeks of age. This means that in many countries, by approximately 72 weeks of age, flocks are considered economically unviable and are slaughtered after approximately 12 months of egg production. This is despite the fact that chickens will naturally live for 6 or more years. In some countries, hens are 'force molted' to re-invigorate egg-laying. This practice is performed on a large commercial scale by artificially provoking a complete flock of hens to molt simultaneously. This is usually achieved by withdrawal of feed for 7-14 days which has the effect of allowing the hen's reproductive tracts to regress and rejuvenate. After a molt, the hen's production rate usually peaks slightly below the previous peak rate and egg quality is improved. In the UK, the Department for Environment, Food and Rural Affairs states 'In no circumstances may birds be induced to moult by withholding feed and water.' Sadly, this is not the case in all countries however.

Other practices in chicken farming include 'beak trimming', this involves cutting the hen's beak when they are born, to reduce the damaging effects of aggression, feather pecking and cannibalism. Scientific studies have shown that such practices are likely to cause both acute and chronic pain though, as the beak is a complex, functional organ with an extensive nervous supply. Behavioural evidence of pain after beak trimming in layer hen chicks has been based on the observed reduction in pecking behaviour, reduced activity and social behaviour, and increased sleep duration. Modern egg laying breeds also frequently suffer from osteoporosis which results in the chicken's skeletal system being weakened. During egg production, large amounts of calcium are transferred from bones to create egg-shell. Although dietary calcium levels are adequate, absorption of dietary calcium is not always sufficient, given the intensity of production, to fully replenish bone calcium. This can lead to increases in bone breakages, particularly when the hens are being removed from cages at the end of laying.

The majority of hens in many countries are reared in battery cages, although the European Union Council Directive 1999/74/EC has banned the conventional battery cage in EU states from January 2012. These are small cages, usually made of metal in modern systems, housing 3 to 8 hens. The walls are made of either solid metal or mesh, and the floor is sloped wire mesh to allow the faeces to drop through and eggs to roll onto an egg-collecting conveyor belt. Water is usually provided by overhead nipple systems, and food in a trough along the front of the cage replenished at regular intervals by a mechanical chain. The cages are arranged in long rows as multiple tiers, often with cages back-to-back (hence the term 'battery cage'). Within a single shed, there may be several floors contain battery cages meaning that a single shed may contain many tens of thousands of hens. In response to tightened legislation, development of prototype commercial furnished cage systems began in the 1980s. Furnished cages, sometimes called 'enriched' or 'modified' cages, are cages for egg laying hens which have been designed to overcome some of the welfare concerns of battery cages whilst retaining their economic and husbandry advantages, and also provide some of the welfare advantages of non-cage systems.

Many design features of furnished cages have been incorporated because research in animal welfare science has shown them to be of benefit to the hens. In the UK, the Defra 'Code for the Welfare of Laying Hens' states furnished cages should provide at least 750 cm^2 of cage area per hen, 600 cm^2 of which should be usable; the height of the cage other than that above the usable area should be at least 20 cm at every point and no cage should have a total area that is less than 2000 cm^2. In addition, furnished cages should provide a nest, litter such that pecking and scratching are possible, appropriate perches allowing at least 15 cm per hen, a claw-shortening device, and a feed trough which may be used without restriction providing 12 cm per hen. The practice of chicken farming continues to be a much debated area, and it is hoped that in this increasingly globalised and environmentally aware age, the inhumane side of chicken farming will cease. There are many thousands of chicken farms (and individual keepers) that treat their chickens with the requisite care and attention, and thankfully, these numbers are increasing.

Contents

Domestic Geese and Ducks. Paul Ives..*page* 1

Ducks and Duck Breeding. Charles Roscoe....................….....*page* 7

Ducks Breeding Rearing and Management.
Reginald Appleyard...........................….....………..…..………...*page* 73

Natural and Artificial Duck Culture. James Rankin….…....…*page* 76

Establishing a Duck Farm

Much thought should be given to the location of a duck farm before deciding where to start; for its location with relation to markets, type of soil, shade and its proximity to water and the slope, or "lay of the land" may be the deciding factor in whether your duck farming will be a success or failure.

All of these will have direct bearing on health of the ducks, economy of marketing, labor and a general but very important bearing on the success of your enterprise.

Consult the County Agricultural Agent of the U. S. Agricultural Extension Service, who, in any section where ducks are raised, you will find well posted and a source of valuable information. He has had occasion to see the best results as well as the failures of men who make duck raising their business and he will be more than glad to pass this valuable information on to you.

The practically universal use of the Pekin as a meat duck on farms where market ducks are the object, would seem to indicate that this is the best choice for raising ducks for meat. If egg production is the object, the White Indian Runner is probably the best choice. The premium on white feathers, a valuable by-product of the duck farm would be enough to turn the scale in favor of a white breed if one were in doubt.

Ducks do best in a temperate climate. They do not mind cold although in extremely cold climates it will cost more for feed and for housing and brooding. But ducks do not do as well in extremely hot weather and so the temperate climate is most suitable.

A light sandy soil is best from a sanitary standpoint. This type of soil will purify itself more quickly than will heavier land, by rains and snow and so help to keep down disease taints and infection. A never failing water supply is very essential. If land can be secured through which runs a sizable stream with land on either side on which yards can be run down to the water, it is very desirable and will save much labor in the carting or piping of water and will also add to the health and vigor of the breeding ducks. If the land is sloping or rolling, so much the better, as this will help to drain and purify the runs without labor of ploughing or tilling.

The location with relation to markets is important. The nearer it is to a good market and a railroad station, the less time and expense will be consumed in transporting the dressed ducks or eggs from farm to market and grain, litter and other supplies to the farm. Not only the distance to market is important but the kind of road over which produce must be transported. If possible, locate on a state highway or one which is kept cleared of snow and which is kept always in good condition. While the local market may absorb most of your product there may be times when there will be an over supply and so it is wise not to locate your farm too far from one of the great cities where surpluses may be shipped without too much delay or length of time in shipment.

Do not depend on salt or even brackish streams for the ducks to swim in even if fresh water is supplied them; they will persist in getting some of the brackish water and while this may not actually injure them, it may hurt their appetites and lower the food consumption to the point of impaired growth; and quick

Dressed duckling. The main feathers of the tail and wings and the feathers of the neck part of the way from the head to the body are left on. The rest of the body is picked clean. (Photograph from the Bureau of Animal Industry, U. S. Department of Agriculture.)

growth through heavy and efficient consumption of feed is the most important factor in growing ducks for meat.

On the scale of business planned, depends largely the type of buildings required. In fact, if only a few ducks are to be raised, there will probably be buildings on the place such as barns, sheds or chicken houses that can, through a little remodelling be used very satisfactorily; or small colony type or shed roof chicken houses may be built. However, if the business is to be on a large scale, more complete and well planned buildings should be planned and constructed.

There must be buildings for laying and growing stock, incubator cellar, brooder houses, killing house, storage and mixing rooms for feed, supplies, etc. One story buildings are more suitable for ducks as they must have easy and convenient access to land. Before planning your buildings, visit other duck farms in the vicinity and consult your County Agricultural Agent.

That the sun may shine into the duck house for its sanitizing and drying effect, it is well to have it face the south although it is not as important as with hens from the standpoint of the ducks as they stay in the house very little except at night.

The floors in duck houses are usually of dirt but cement floors are better as they may be so much more easily cleaned and kept sanitary. Either way, the floor should be about a foot above the level of the ground and well covered with sand or litter. Shavings, sugar-cane litter or straw are materials generally used and there is little to choose between them. Litter in cold weather may be allowed to accumulate as the volume under foot will add to warmth but care must be taken to keep it reasonably dry and for this reason all the feeding and watering should be done on the outside of the house if that is possible. New litter is

ESTABLISHING A DUCK FARM

added as frequently as needed, depending on the weather and condition of the old litter.

When planning floor space in buildings, figure from three to four square feet per bird. The greater number together, the less feet of floor space for each bird. It is a question how many to run in one flock. Any number from 10 to 500 are run in flocks but usually from 100 to 150 is customary. If no more than this number are run together, there will be less cripples and fewer broken eggs but it is more expensive in building and labor to house them in the small units, where large numbers running into the thousands are kept.

Many duck growers use no nests at all but as a rule nests are provided on the floor level around the outside of the room against the walls. They should be about 12 inches wide in front, 15 to 18 inches deep with partitions a foot high. The partitions may be held in place by a 5" by 7/8" strip set on edge in front which forms the front of the nests, the wall of the building

On this plant, the lay of the land was such that not all of the yards could be run down to the stream. So a shallow canal was dug from the stream through the yards which were without natural water frontage. (Photograph from the Bureau of Animal Industry, U. S. Department of Agriculture.)

POSSIBILITIES OF THE DUCK

Ducks for the Backyarder

By " backyarders " is meant those persons who are obliged to keep their birds confined within the limits of a very small garden or backyard. Although one frequently hears it said that ducks are entirely unsuited for backyard laying, nevertheless, the writer has come across many actual cases where ducks are, or have been, kept under the most restricted conditions with the greatest of success. For instance, some years ago the writer was invited by a woman to visit her backyard poultry farm in a side street off Clerkenwell Road, which is in the heart of London. In another case a local milkman informed me that he was keeping five Khaki Campbell ducks in a wash-house attached to his cottage, and that they produced eggs regularly. The writer knows of many other instances where ducks are being kept at a profit, and in a healthy condition, in strict confinement. It may be said that to do so is to inflict grave discomfort and misery on the birds, but this cannot be so, for ducks are very sensitive creatures, and, if unhappy, nervous or badly fed, they will not lay eggs. The mere fact that they lay well is an indication of health and happiness.

However, if ducks are to be kept in close confinement, or within a very restricted space of ground, strict cleanliness and sanitation must be observed.

The floor space should either be cemented, or tarred in the case of an earth floor, but preferably the former. If this is not done, great difficulty will be experienced in keeping the surface from smelling. Apart from the unpleasantness to oneself and family, who may be willing to put up with it, you can take it from me that your neighbours will not, nor will the Sanitary Inspector, and you will find yourself in serious trouble with the local health authorities.

Providing, however, that the floor space is covered with a material that can be regularly washed down, the dirty water drained away, and some chloride of lime is sprinkled into or around the drain or soakaway afterwards, the ground can be kept clean and sweet, and there will be no complaints as to smells.

Apart from smells, you must not allow your neighbours to be annoyed by noise from the ducks. That, too, is an offence against local laws. If the ducks quack a lot, remember that you, and not your neighbours, are wrong, and, furthermore, you cannot be looking after your ducks properly, for if sufficiently fed and fed at regular intervals, they will not quack, except, perhaps, if they are frightened. If your neighbours complain, don't " go off the deep end " and start a row, or want to go to law, but talk to them nicely, even if their language and tone are somewhat heated. Be reasonable, for " a soft answer turneth away wrath," and assure them that you will do all you can to rectify the nuisance—and do so ; don't just talk about it. Half the troubles in this world are brought about by persons being unreasonable with one another.

I make no excuse for this little dissertation. Apart from keeping ducks, I have spent many years working as an official in the Law Courts, and have listened to many actions between poultry-keepers and their neighbours. Actions which have cost one side or the other, or both, a great deal of money, which might have been saved had one or both exercised a little forbearance and carried out that priceless axiom, " do unto others as you would that they would do unto you."

Now for a few more words about the space required for intensive duck-keeping. Providing the floor is covered with a material that can be washed down, then allow five feet of floor space per duck. If you have more room to spare, so much the better, because five feet per duck is the minimum.

The maximum number of ducks to keep under actual backyard conditions is half a dozen. Exceptional circumstances and exceptional people might call for an alteration in this rule, but owing to the high standard of cleanliness required in backyard duck-keeping, I cannot visualize the average person conforming with this standard. Besides, the bigger the number kept, the worse the smell, and the bigger the noise that will come forth from them if you fail in your duty to the ducks and your neighbours.

It seems to me that backyard duck-keeping should be confined to producing just enough eggs to satisfy family requirements, and not to attempt to run them as a commercial proposition. Keep just a few of the best ducks you can afford to buy under the best possible conditions, rather than a larger number of poorer quality ones kept poorly. The former will probably pay you much better. You will become greatly interested in them and proud of them, whereas in the latter case, you will keep quiet about them, and certainly lose interest in them when they prove a " wash-out."

Ducks are wonderful creatures, and the modern highly prolific Khaki Campbell duck is undoubtedly the most efficient egg-producing machine in the world. To get the best results from them, make pets of them, let them become completely tame. Feed and treat them very well, and I can assure you that you will be more than repaid for your trouble and affection. Backyarders have written to me, telling me of phenomenal egg returns, records that no-one could believe possible unless they had kept this breed of duck. One correspondent in Manchester sent me the record of five Khaki Campbells kept in his backyard, and each of these ducks had exceeded 300 eggs in twelve months, and one had laid 327 eggs. Think what that means in money ! Even at a penny each per egg, this represents over £6 from five ducks ! You can start with two ducks if you like—two really good ones. They would cost you about 10s. each, if purchased when three months old, or half that price at six weeks. A sugar box would house them, and a small run, consisting of three more sugar boxes with the end boards removed and placed end to end, would suffice at a pinch. There is no necessity for elaborate housing for ducks. They will lay just as well sleeping in a sugar box as in a marble hall.

Scraps from the house, a little boiled cabbage, and meal added, would feed two ducks, and they would turn this crude food into a couple of good-sized eggs day after day with almost unfailing regularity, until you will just become fascinated by them, and your wife will bless you for having a useful hobby.

For further details relating to suggested breeds to keep, method of housing and feeding, see Chapters II, V and VI.

Ducks for the Small Poultry-keeper and Holder

It is to the smallholder, the man with an allotment or piece of garden or rough ground to spare, that ducks hold out the greatest possibilities. Providing good stock is purchased, and the birds are carefully tended, they will most certainly represent the most profitable line that the holder of the ground can possibly cultivate.

Official laying trials have amply demonstrated that the laying type of duck, kept in small runs, give remarkably high egg returns. In fact, the results have been so encouraging, that the writer regrets that laying tests on the small flock lines were not continued. As a practical demonstration of the capabilities of the laying duck, these tests were of the greatest value to those smallholders who were anxious to get authoritative information upon the subject.

It was in 1922, at the National Laying Test at Bentley, that the first test of small flocks of ducks was made; actually, they consisted of pens of five ducks, confined in a run of 150 square yards. The average for all breeds was 204.65 eggs per bird, as against 188.25 eggs per bird for the ducks run in the large flock, but these scores were easily beaten later on.

Although in the laying tests only five ducks were kept in each pen of 150 square yards, the writer considers that it would be safe to keep at least eight ducks permanently on this area—providing they were given a certain amount of green food, in addition to any they might find on the run itself. Further, if double the space could be allotted to ducks, say 20 yards and 15 yards, and a small portion of the run nearest the entrance were cemented over, a patch 10 feet by 4 feet would suffice, and the ducks were fed and watered there. It would then be possible to keep fifteen of the lighter breeds permanently confined to a run of these dimensions.

Rearing Table Ducklings

The foregoing remarks apply to cases where ducks for egg production would be kept, but the same area of land could very well be used for rearing to killing stage a very considerable number of Aylesbury or other breeds of table ducklings, as well as maintaining a breeding pen of Aylesburys. In the latter case, it would mean wiring off sections of the run for rearing purposes, but during all the autumn and winter months the breeding pen could have access to the whole run if necessary. Wiring off is a very simple and cheap business in the case of Aylesburys, for 12 to 18 inches high netting will suffice.

The sketch on page 10 shows a suggested layout for semi-intensive rearing, but to-day the tendency is to rear ducklings to killing stage intensively, that is, to keep them in confinement, rearing them in large wire netting cages, which is quite a successful method, and enables the owner to rear very large numbers upon an extremely confined space. As it is, even on the semi-intensive method, 3-4,000 ducklings can be reared in a season, that is, between January to end of September, upon $\frac{3}{4}$ to 1 acre of ground, not merely for one year, but year after year, providing the soil is of a light gravelly or sandy nature.

In the sketch it will be noticed that the ground is divided into five sections. After March the stock birds would not require any housing accommodation, unless the weather was very cold, so that their winter

POSSIBILITIES OF THE DUCK

quarters might well be used as the brooder house for the youngsters. The earliest ducklings being hen-hatched would be reared by their foster-mothers.

A makeshift hutch, for the stock birds to lay in, can always be rigged up out of a couple of sheets of corrugated iron as a roof, and a few sacks for sides and some straw as litter. This should not be necessary, however, unless late frosts and snow are being experienced. Hatching eggs should not be allowed to get frozen, hence some form of protection in cold weather.

The reason for my suggesting that the run should be divided into five sections, is because it is far better to hatch out a couple of dozen ducklings each week, than to hatch 100 one week and then no more for a month, the reason being that unless a large batch of ducklings is brought off very early in the season, when prices are high, and can be disposed of at market at profitable prices, later on it would be found that 50 ducklings, all hatched at the same time and therefore due to be killed on the same day, would be by no means so easy to get rid of at the best price. London and Provincial salesmen have regular wholesale customers, whom they know will take so many ducklings a week, but are not likely to buy more unless the price happens to be lower than usual. If, then, a casual customer suddenly sends along fifty ducklings one day, Mr. Salesman may not be able to place them at once, and any delay would cause a serious depreciation in price. On the other hand, if the fifty had been spread over four weeks, and the salesman notified that there were regular weekly lots of a dozen coming along, he could then find a regular customer for this lot, and there would be no risk of the birds being left on the salesman's hands for a day or so.

This being so, having decided to go in for table duckling rearing, it is essential to aim at regular weekly supplies of ducklings, and essential, too, to stick to a certain day or days for killing and delivering them.

SUGGESTED LAY-OUT FOR SMALL DUCK REARING PLANT TO PRODUCE TWENTY FATTED DUCKLINGS PER WEEK BETWEEN FEBRUARY AND OCTOBER.

1. Food and plucking shed.
2. Rough shelter for hatching hens.
3. House for stock ducks.
4. Duckling house.
5. Duckling house.
6. Duckling shelter.
7. Cemented feeding patch.
8. Shelter.
9. Cemented feeding patch.
10. Bathing pool. 6 ft. diameter.
11. Shelter.
~~~ Denotes fir or laurel hedge as wind breaks.

POSSIBILITIES OF THE DUCK

Especially is this the case where the intention is to market them in London or one of the large provincial town markets.

Here, now, are the rough pre-war costings and estimated profits possible on the production and sale of 400 good quality Aylesbury ducklings.

Rearing Costs

In the case of a smallholder, who rears 400 Aylesbury ducklings in a season from his own breeding stock of eight ducks and two drakes, although it would be possible to rear rather more than this number with luck, the cost of rearing would work out somewhere near these figures:—

	£	s.	d.
Cost in September of 8 May hatched ducks, at 8s. each	3	4	0
Cost in September of 2 May hatched drakes, at 8s. each		16	0
Cost of feeding 10 stock ducks for 12 months, at 16s. each	8	0	0
Cost of rearing 400 ducklings to 10 weeks, at 2s. 6d. each	50	0	0
Cost of packing and transport, at 6d. per duckling	10	0	0
Cost of labour, at 1s. per duckling	20	0	0
	£92	0	0
Sale of 400 ducklings, average price 6s. each	120	0	0
Sale of 10 stock ducks, average price 4s. each	2	0	0
	£122	0	0

Showing a net profit of approximately £30 or, roughly, 1s. 6d. per bird. This may not sound much, but it must be remembered that the profit is made in a very short time. Further, I have allowed a liberal amount for labour and rearing expenses. My own experience of rearing, not counting my own time, is that it is possible to clear £12 net per hundred ducklings reared and sold. This figure is not far from the actual mark, and might be exceeded if one hatched a good percentage of early ducklings, when profits are very much higher.

The actual cost of feeding an Aylesbury duckling, from one day to ten weeks, should not be more than 2s. 6d. in these days. In the costings given, allowance has been made at the rate of 9d. per duckling for loss on eggs from the mother hens, litter, depreciation on appliances, etc.

The prices for both feeding stock birds and ducklings have been based on a feeding mash costing 9s. per cwt. (present prices are higher than this). In order not to under-estimate costs, should the value of feeding stuffs rise beyond present prices, which they are almost bound to do, and unless the price for fattened ducklings rises in like proportion, the net profit would be smaller. Actually, the season in 1939 showed reduced prices for ducklings over the previous year. This was probably due to mass production of a medium weight white duckling, weighing about 4 lbs., by certain firms. Such market conditions do arise from time to time, but generally rectify themselves.

Ducks as an Adjunct to the General Farm

In 1920, in conjunction with a friend who owned a 500-acre general farm, I decided to try out the scheme of running a large flock of Khaki Campbell ducks upon some 115 acres of low-lying water meadows, which for a considerable part of the year were more or less floating meadows. Through this section of the farm ran watercress beds and many ditches and dykes. The land appeared admirably suited to ducks, far more so than for cattle, as there were many casualties from Johanne's Disease amongst them when they had had a spell on this low-lying ground.

POSSIBILITIES OF THE DUCK

To commence with, I provided a couple of dozen ducks and four drakes. A couple of Hearson 120 egg size incubators were purchased, after trying two others of a cheaper variety that proved quite hopelessly inefficient. From these few stock birds, and by purchasing a hundred day-olds, a flock of 300 ducks was ready. Incidentally, it was decided that it would be better to run some drakes in with the ducks, in order to prevent them wandering away with wild ducks which visited the meadows in considerable numbers.

At any rate, the results from the flock of 300 exceeded all expectations. They commenced to lay in September, the majority having been hatched out in May and June. By the end of October the output touched fifteen dozen eggs per day.

The labour involved in connection with the maintenance of the ducks were very little. At night the birds had a mash consisting of flaked maize, wheatings, and coarsely ground wheat and oats, plus 10 per cent best quality fish meal. The food was put into a number of metal troughs standing in the stock yard. Any food left over, after about two hours, was emptied into buckets and kept over towards next morning's breakfast, which was served at about 7 to 8 a.m. winter and summer. It does not matter about it being dark, for ducks see quite well in darkness.

The quantity of food consumed per head varied widely, according to the season of the year. In early spring, and on to July, the ducks found quite 75 per cent of their food, often not touching their evening meal. In winter the consumption averaged about 6 ozs. per head. The ducks were always given as much food as they could eat, for only by so doing can one get the maximum egg output from them.

The greatest amount of labour involved was in connection with washing and packing the eggs; generally one of the womenfolk of the household did this job. As soon as collected, the eggs were put into large milk pans and warm water poured over them. After remaining to soak for ten minutes, they were thoroughly wiped, rinsed, and laid out to dry in slatted-bottomed trays.

The egg yield continued to increase, until by mid-January of the following year the ducks were producing in value of eggs more money than did twenty-one milking cows. The weather happened to be mild, and natural food abundant in the form of frogs, slugs, etc., which made their appearance and were consumed in large numbers by the ducks.

The birds were kept at night in a small stock yard attached to some piggeries which were empty, and the birds had access to the sties. The whole yard which was roughly 30 feet by 60 feet, was littered down with rough straw. As the straw became dirty and wet, more was put in, until eventually it became a couple of feet deep. The yard was then mucked out.

By April of the same year, on quite a number of days, the output of eggs exceeded the number of ducks. This may sound impossible to those unacquainted with the laying of Khaki Campbells, but it is quite a common occurrence for a duck to lay two, sometimes three, eggs within a twenty-four hours day during the flush period. The writer having trap-nested many ducks, has frequently found two eggs in the morning laid by a particular bird within a period of twelve hours. These eggs have been properly shelled. Occasionally, the same bird will lay a third egg, but without a shell.

My friend was so enthusiastic about the results obtained from this flock of ducks which, by the end of their first laying year, had averaged 214 eggs each, and had shown a net profit of just on 11s. per head, that he decided to increase their number to 500. Further, it was considered worth while to run on all the second season ducks for another year, so that we only had to breed another 200 pullet ducks. This was easily done, the eggs being obtained from the original flock, which, in spite of the fact that there were only twenty drakes running with 300 ducks, were remarkably fertile and hatched very well in incubators. Fertility was generally 100 per cent, and hatches averaged 85 per cent. The ducklings were very strong, and were reared with ease.

The whole outlay on apparatus did not amount to more than £50. £30 of this sum was for incubators, the rest was spent on hoovers and small lightly made wooden houses for rearing the ducklings in.

Later experience went to show that it is best not to exceed flocks of over 200 ducks, unless one divides the birds up and spreads them over the land at one's disposal. It was found that the majority of the ducks would not range far enough to take advantage of the natural food which the ground could provide them with, consequently, as the ground around the farm became depleted of insects and frogs, so the egg yield diminished, unless more artificial food was provided for them. This, of course, increased the cost of production.

When the large flock of ducks first began to range over this water-logged land, the ponds and ditches were alive with insects and frogs. In a year's time there was not a frog to be found within 500 yards of the farm. The ditches and ponds became cleaned of weeds, and the ducks began to attack the watercress beds, which eventually had to be wired off. To give some idea of what 300 ducks can consume in the way of natural animal food, I weighed up some worms, and found that there were about seventeen good-sized earthworms to an ounce. A large frog weighed $1\frac{1}{2}$ ozs. A full-grown Khaki Campbell duck in a day can, and will, consume seven ounces of food; if possible, she would prefer to eat almost entirely natural animal food. This means that she could consume two large frogs and sixty-eight large earthworms a day, so that in three months 300 ducks could dispose of just on 55,000 frogs and nearly 2,000,000 earthworms. It is therefore quite obvious that, unless moved about, this flock of ducks would completely clear earthworms, frogs and slugs off a large area of land in the course of a year.

For those who care to work out the problem, Darwin calculated that in ordinarily good soil, there were thirty-four earthworms per cubic yard of soil. If correct, my rough calculations seem to arrive at the fact that in twelve months, 300 ducks could, if they were able to get hold of them, consume all the worm inhabitants over an area of thirty acres, but that is not taking into consideration the fact that over that period, there would always be breeding going on amongst worm survivors. Furthermore, nature generally has a way of coping with such a position. Nevertheless, it is obvious, and it most certainly was so to us, that part of the land was becoming very much depleted of insect and frog life.

The flock was eventually increased to 700, but the results were not nearly so satisfactory, and the profits per head decreased as numbers increased.

The experiment was not continued, as my friend sold his farm, and the new owner was not interested in ducks, and I found a buyer for most of them. Two hundred of them were sent to a man in Kent to run in a large orchard, and they gave most satisfactory results.

At any rate, the experiment went to show that many general farmers could quite profitably keep a flock of laying ducks, and I know from the experience of customers and correspondents, that where a farmer has a milk round, he can sell his duck eggs without difficulty and make a handsome profit out of them.

Table Duckling Production

If a farmer does not wish to keep ducks for egg production, it is well worth his while to consider the possibilities of keeping a couple of dozen good Aylesbury stock birds, for these can be made to bring in a very nice little harvest in spring time, either by rearing early ducklings, or else by selling settings of eggs or day-old ducklings. There is always a market for good stock.

The cost of starting with first-class stock need be only a matter of a few shillings ; hatching eggs, from a really reliable breeder, would cost about 7s. 6d. to 9s. for a dozen. Then, later, when the ducklings from these eggs were full grown, a stock drake of a different strain should be purchased, and, again, it is infinitely better to purchase one really good one for, say, 10s. to 15s., than a couple of indifferent ones at 6s. each.

From this small start, it would be easy to work up a small and selected flock of breeders, which on a general farm could be run very inexpensively and profitably. A Cornish farmer, whose wife accommodated visitors during the summer months, told me that during the holiday season he could dispose of nearly 200 fat ducklings, and that they made good prices ; and he reckoned that, considering the small amount of capital outlay expended on them, and the very small amount of labour required to look after them, he cleared very nearly the same amount of profit per head on them as he did on the sale of his lambs.

Ducks as a Sole Commercial Undertaking

Ten years ago I would have replied, " Yes, certainly," to a query as to whether a man could make a living out of keeping either laying or table ducks upon a commercial scale. As conditions stand to-day in 1939, it is doubtful whether the ordinary person could make a duck-farm pay. A few exceptional persons do so, and others could if it were done on a sufficiently large scale, upon suitable ground and close to a good marketing centre. But even then the proprietor would have to be a good business man, for the profit margin is small, both on eggs and table ducklings.

The person undertaking a commercial duck-farm to-day must be energetic and enterprising. He must not lay out an unnecessary amount of capital on dead stock, that is, upon expensive houses, mammoth incubators, and complicated brooding appliances. He must possess experience of ducks, for there is little margin for mistakes, and he must have at least £500 capital.

The kind of situation that I would seek for a duck-farm would be a good acreage of marshy land near a big city, land that could be acquired or rented very cheaply, and which would afford a good natural feeding ground for about 300-400 laying ducks; that would mean somewhere around 50 acres.

Laying ducks alone, however, would be insufficient and too uncertain to produce a regular income, and it would be inadvisable to put all one's eggs into one basket by devoting the whole business to the production of duck eggs for the table. Of latter years, the late spring and summer prices have been exceedingly low.

On the other hand, if a considerable number of table ducks were reared during these lean months, the average monthly earnings could be kept up and a useful profit shown on the year's turnover.

I suggest that a flock of 300 laying ducks should be aimed at. These should be fed for high autumn and winter egg production, and then, when the time of low prices comes along, these ducks should be put on to a mere maintenance ration of potatoes or maize, the rest of their food being obtained from this free range. When nearly through their summer moult, they would be fed up again in preparation for autumn and winter egg production, for one November egg is worth two in April—and it is not difficult to get up to 60-70 per cent production in autumn with a flock consisting of half first season and half second season ducks.

The natural food obtained from the land plus, say, three ounces of boiled maize or six ounces of boiled potatoes per day, should give quite a satisfactory egg yield between March and July. A very high spring and summer egg production, which can be obtained by heavy feeding, generally tells its tale during the following autumn. The ducks become so depleted in body, fat and stamina, that they are unable to withstand the strain of heavy laying during the cold weather; consequently, they fall into a second moult in autumn. A moderate spring production assists in avoiding this out-of-season moult.

As regards the table duck section, here again the situation has considerably changed, even within the last three years. The tendency now is to rear a crossbred white duckling, weighing about 4-4½ lbs. dead weight, rather than the higher class large Aylesbury duckling running up to 7 lbs. at ten weeks old. These small ducks are generally from an Aylesbury drake x white Campbell-Aylesbury cross duck. This produces a useful table bird, and the breeders are better layers than pure Aylesburys, and much more likely to lay eggs during the autumn and winter than are pure Aylesburys. The result is that these smaller white ducks can be produced at a cheaper rate. The average good stock Aylesbury duck produces on an average 40-50 rearable ducklings in a season. She does not hatch the eggs herself, of course, but they are hatched either under hens or in incubators.

A good white Campbell x Aylesbury duck should be able to produce nearly double that number at a lower cost of feeding for the stock bird and for the fatted ducklings.

The 1939 market price for the crossbred Aylesbury duckling has been only 1s. to 6d. under that of a much larger Aylesbury duckling. The difference in the cost of production between the two is about 1s. 6d. per head in favour of the crossbred, and the profit on each very nearly the same. It may be that the pure bred Aylesbury will come into its own again, but as we must be practical, and I am dealing now with the matter from an entirely commercial point of view, the £ s. d. side of the question must have first consideration, and sentiment must go by the board.

In a subsequent chapter dealing with the various breeds of ducks, I have fully dealt with the merits of each variety, and under Aylesbury ducks, a considerable amount of space has been devoted to this outstanding and purely English breed of table duck.

In the opening of this chapter, the sum of £500 has been mentioned as the minimum amount of capital required, whereas in the booklet on "Ducks and Geese," published by H.M. Stationery Office on behalf of the Ministry of Agriculture, the author of this book mentions a sum less than half my suggested minimum capital requirement. The difference, however, is due to the fact that on the smaller capital, an estimated profit is shown which would be quite insufficient to keep a man and his family, or for keeping the man and his family until such time as the proposition becomes a paying one. My own view, after years of experience, is that it is essential to have at least a year's capital reserve for cost of living. To start any business, with the idea that it is going to keep one from the word "go," is to court trouble. Where rearing poultry is concerned, one is, if not very experienced, almost certain to make expensive mistakes, which cost hard cash, and one really bad mistake might well put a small man out of business, had he no capital reserve to fall back on.

At present-day prices, big profits on small turnover of commercial poultry are out of the question, and, as before stated, there is little or no margin for errors. There used to be a very profitable margin, and there may be possibilities in the future of making much better profits than can be made to-day, but I am not prepared even to guess at what may happen. The whole object of this book is to give practical advice, and to come down to "brass tacks," and to put off rather than to encourage the wrong type of person from going in for a business in which he would most certainly lose his capital and of which he would make a failure.

Just to quote one most unfortunate case. Some years ago a man came to me, and said, "I am absolutely penniless as the result of reading one of your articles on commercial duck-keeping. I became enthused and went in for it on a big scale, and was bankrupt in a year and a half." He then told me how he had spent a lot of capital on houses, wiring and apparatus, and a number of expensive stock birds, and by the end of a year had a large flock of laying ducks which wouldn't lay, or, at any rate, would not lay sufficiently well to show a profit, and after another six months he was hopelessly in debt and had to sell up.

I pointed out to him that in the article I had written, it was particularly stressed that it was advisable to start with a small number of birds, gain one's experience, and then work up the number. Also, that the whole theme of the article was the necessity of having the right conditions and the avoidance of laying out money on expensive houses and apparatus. He had completely ignored these essential points, and without practical experience had attempted to start on a considerably larger scale than I would have ever advised or had even experience of.

I have knowledge of others who have done exactly the same thing, but not on quite such a large scale, but with the same unfortunate results. They were all persons who refused to be guided by the old, old maxim of " Don't try to run before you have learned to walk."

Duck-farming, like any other form of farming, will have its good and its bad years, and only by averaging out the profits over a series of years can one really arrive at a true idea of the possible income to be derived from a given number of ducks kept correctly, and the produce marketed efficiently and with businesslike ability.

THE BREED OF DUCK TO KEEP

As egg producers. For table purposes. As a dual-purpose bird. For ornamental purposes.

As Egg Producers

When it comes to deciding which breed of duck to keep for the purpose of egg production only, our choice is narrowed down to seven possible varieties. Of these, there are three outstanding breeds, and one of them, for the last eighteen years, has proved itself at public laying tests to be the paramount laying breed.

Taking them in order of merit, they are as follows :—
(1) Khaki Campbells ; (2) Fawn and White Runners ; (3) White Runners ; (4) White Campbell ; (5) Buff Orpingtons ; (6) Fawn Runners ; (7) Magpies.

The Khaki Campbell duck represents the White Wyandotte or Rhode Island Red of the chicken world. The White and the Fawn and White Runners are equivalent to the White and the Black Leghorns.

If we take the laying test results over a long period of years, we find that in the English tests where a duck section has been included, Khaki Campbells have shown the best results in 90 per cent of them, and in the post-war National Laying Trials held at Bentley, which were of international importance, and when the duck sections were very strongly supported, four out of five of them were won by Khaki Campbells, and some phenomenal scores were piled up by them.

Looking back at my notes on the duck sections of the National Laying Tests, I see that in the 1922-3 test Khaki Campbells easily came out top with an average of 211.25 eggs per bird in the large flock section, and 227.73 eggs per bird in the small flock section. Indian Runners produced 171.56 and 194.88 respectively in the same sections. The following year Khakis again were winners, although there were some excellent individual scores by Runner ducks. As a matter of fact, individual pens of White and Fawn and White Runners, and individual birds of these breeds, have and do put up some astounding records. It is the average production per bird that counts, however, and it is in this respect that the Khaki Campbell has scored so highly.

When the National Laying Test ceased to run a duck section, the Harper Adams Agricultural College Laying Tests continued the good work, and we see that year by year the average output per Khaki Campbell duck improved, whereas that for the Runner breeds appeared to remain about the same.

In order to give readers some comparative idea of the respective laying powers of the breeds represented at the 1926-27 and 1927-28 duck sections of the above tests, here are the average scores for these two years :—

Breed	Numbers	Egg Average 1928	1926-27
Khaki Campbells	96	242.7	231.47
White Runners	18	203.6	191.5
Fawn and White Runners	18	206.7	209.9
Buff Orpingtons	6	204.0	—

In the single pen tests for individual birds :—

Khaki Campbells	14	263.7	258.1
Other breeds	3	190.6	226.7

Two Khaki Campbells, during the 1927-28 test, put up the following scores in 365 days :—

Duck No. 1340—357 eggs, weighing in all 901.6 ozs.

Duck No. 6—356 eggs, weighing in all 862.1 ozs.

The first of these two ducks laid all special and first-grade eggs.

The cost of food per duck for all breeds averaged out at 1s. 4d. per month throughout the tests for an average production of 19.2 eggs per month, and this showed a margin of 14.05d. per month over cost of feeding. At 1939 prices, the margin would be considerably lower, probably just about half, say 7d. per month or 11s. per head per annum, taking the average production at 220 eggs each. This average of 220 eggs per duck is only half a dozen more than I have obtained from much larger numbers of commercial **Khaki Campbells**, kept under ordinary farm conditions, with no attempt at selecting for outstanding results.

Incidentally, at the afore-mentioned duck laying tests the average weight per egg for 299 birds was :—

Khaki Campbells	2.51 ozs.
White Runners	2.54 ,,
Fawn and White Runners	2.65 ,,
Buff Orpingtons	2.60 ,,

It will be noted that these egg weights all exceed the weight for " specials " for hen eggs.

There has been little variation in the highest production of Khaki Campbells at laying tests in the last few years. The best pen averages were somewhat lower, if anything, in 1938, than they were ten years ago.

At the first duck laying test in Scotland, where the egg scores were generally lower, eighty Khaki Campbells averaged 160.6 eggs each, as against 146.98 per bird for Runner ducks entered in the trial, perhaps not a great difference, but represents at to-day's prices about 1s. 6d. per head greater earnings in favour of the Khaki Campbell.

My own experience when trap-nesting Khaki Campbells is, that if one starts building up a flock from a pen of really good trap-nested birds, possessing good-sized bodies and typical of the breed that they are supposed to represent, with records around 230 eggs each in their first season's laying, making sure that the eggs are of good texture and weight, not less than $2\frac{1}{4}$ ozs. and not exceeding $2\frac{3}{4}$ ozs., such a pen will produce progeny equal to themselves and often much better, depending to a great extent upon the quality of the drake mated to these birds. The drake should come from really choice parents, trap-nested for several generations. Parents that have shown not only high laying records, but good records for health, physique, for fertility and hatchability of their eggs, and good rearing qualities of their young. All of these are most important points, because one must never lose sight of the fact that upon the quality of the drake depends, to a very great extent, the laying qualities of the female progeny coming from the ducks to whom he is mated. In other words, it is the blending that counts so much in mating results.

As an instance of how great an influence the drake can have upon the laying qualities of his progeny, I mated five good ducks to a good-looking drake, whose pedigree I knew nothing about. From this mating I hatched some twenty ducklings. The drake was then taken out, and after a fortnight another drake was mated to the same ducks. This latter drake came from a duck which had laid 317 eggs in a big laying test, and had been mated to a drake coming from a very prolific dam. I had seventeen ducklings from this mating. The ducklings from both matings had been toe-punched, that is, holes pierced in the webs of their feet so as to identify them. Later on, I selected two good-looking young ducks from mating No. 1 and four from mating No. 2, and sent them up to the National Laying Test, one of No. 2 being entered in a single duck laying test.

The result was that the two ducks from No. 1 mating let the pen down badly, and put up a very poor show, but so well did the other three lay that for months my pen was second and third in the test, and finally finished fifth; one duck from No. 2 mating laid 314 eggs, another 263, and the third 247. Of the other two coming from No. 1 mating, one only laid 174 and the other 209. The bird in the single pen competition, also from mating No. 2, came out second with 283 eggs.

After the competition, I built up a small flock of 63 ducks from the No. 2 mating. They ran in a grass pen 50 yards by 10 yards, and at the end of twelve months these ducks had averaged just over 264 eggs each. Those that I had bred from No. 1 mating, apart from the ones sent to the test, soon showed themselves as bad starters and erratic layers when they did get going, and were disposed of before the season was through.

Apart from Khaki Campbells, Fawn and White and White Runners, and Buff Orpington ducks, there are quite a number of other varieties which are moderately good layers, but are not worth considering when it comes to deciding which breed one is likely to make most profit from as commercial egg producers.

If one is going to keep ducks upon a commercial scale, few persons can afford to experiment, and it is much better to plump for Khaki Campbells as layers. They have proved their merit over and over again in public tests. It is the average that counts, not the results of a few star layers. Furthermore, Khaki Campbells are rather less temperamental than are pure Runners, and that is important when running a large number of ducks together.

It does not pay one to rear Khaki Campbell drakes for table purposes. It is much better to sell the ducklings as day-olds or when quite young, and then dispose of the drakelets in the best way one can. Kill them off, if necessary, immediately they have been sexed. See Chapter IV for the way to sex ducklings.

The only time when it may pay one to rear a few drakes up to killing stage, is when they are being reared on a general farm where there is a surplus of skimmed milk and plenty of small potatoes (chats) available. A mixture of boiled potatoes, milk, and boiled green stuff, will fatten them pretty well and cheaply. As a matter of fact, the Khaki Campbell drakes make very good eating, as they have a flavour not unlike game, owing to the presence of Mallard blood in the breed.

The Khaki Campbell Duck

It is said that the origin of the breed was a cross between a Rouen drake with Fawn and White Runner ducks, and at some time or other some wild Mallard

THE BREED OF DUCK TO KEEP

blood was introduced. At any rate, the early Khaki Campbells clearly showed the presence of the Fawn and White Runner blood, for few of these Khaki Campbells were without a white bib or some white under their bill. But as to the presence of wild blood, I am doubtful, for the traits attributed to the Mallard are frequently found in Rouen cross-bred ducks, and there is little doubt that the Rouen is a direct descendant of this breed, but generations of selection has brought them to their present form.

In latter years, breeders of Khaki Campbells have paid much more attention to the standard of perfection laid down for this breed than they used to, and, in my opinion, it has been rather overdone. Although agreeing with the policy of breeding Khaki Campbells so that they approximately conform to a standard of markings, the idea of sacrificing in any way their commercial qualities for the sake of such trifling details as evenness of the feather lacing, or for the dark streak which one so frequently sees running from the eye towards the bill, seems absurd.

The Khaki Campbells, when originated, were undoubtedly very near to the colour of Indian Khaki, which is lighter than the present army khaki. In the standard of perfection laid down for this breed of duck, by the Khaki Campbell Duck Club, it says :—

"The colour shall be khaki all over, the ground colour being as even as possible. The back and wings and breast showing even lacing of the same colour, or of as near the same colour as possible."

At one of the last Crystal Palace Poultry Shows held, I sought out the Khaki Campbell classes, and found the winning birds were no longer khaki colour, but a definite dark brown with very defined lacing on the breast, back and sides. They were not of the withered grass shade.

For the benefit of those readers who may wish to exhibit birds, here are the essential points to be observed for Khaki Campbells.

DUCKS AND DUCK BREEDING

THE DRAKE. Head, neck and wing bar bronze, a brown bronze being preferable. Weight 5 lbs.—not more for commercial birds. He should have a fine head, with eyes set well up in the cranium, a bill that is strong yet not stubby or dished, that is, curving downwards in the centre. In a well-balanced bird, the bill and the head are the same length. The neck is an important feature. It should be long and taper slowly to the shoulders. These should be broad and well developed, likewise the abdomen, but not dropped like that of a duck in lay. His body should be long and rather low, better low than too high on the legs. Compactness with gracefulness of line is what one seeks. He should be sprightly in his movements; the sluggish drake does not, as a rule, make a good stock bird.

As regards the correct plumage for a drake of this breed, the head and neck should be a bronze or green shade, with no suspicion of white under the bill, the body being an even shade of warm khaki, with legs and feet a dark orange. The under-wing plumage should be a soft grey (preferably dark grey), but never cream or white.

The bill must be green—an olive to bluish green—with a black bean-shaped marking at the tip. A yellow or yellowish green coloured bill is a serious defect from an exhibition point of view.

In the case of the duck, fine head points are essential. The head and neck should be on even finer lines than that of the drake. The eyes should be set up very high in the skull, and somewhat prominent, bright and intelligent looking. In three months old youngsters the head and neck give one the best indication of all as to the possible prolificness of the bird. A bird with a fine head and snaky neck is rarely disappointing. Extreme fineness sometimes means, too, a rather furtive, nervous temperament, but rather this than a coarse-headed, sluggish duck.

A PEN OF RECORD-BREAKING KHAKI CAMPBELLS

KHAKI CAMPBELLS BEING KEPT INTENSIVELY IN A RAISED WIRE RUN

THE BREED OF DUCK TO KEEP

When the duck is nine months old she should, if well grown, weigh not less than $4\frac{1}{2}$ lbs. She should have well pronounced shoulders, and appear well balanced and even in shape.

THE DUCK. This has been mentioned above, except that on the wings, which have a bar marking across them, a lighter shade of khaki is permissible. The bill should be a greenish black; the legs and feet a dark khaki colour.

Serious defects from an exhibition point of view are a yellow bill, a white bib, white under-wing, yellow legs, white wing bar, and any deformity. It is waste of time and of one's entrance fee to exhibit birds possessing any such defects.

If one wishes to breed birds suitable for showing, one must be very careful about selecting the drakes, because the eye streak is often inherited from the male and this cannot be noticed whilst he has his winter plumage. Select the drake, therefore, after having seen him in his brown summer plumage; and if he has an eye streak, then don't use him.

Now, as regards the physical features most desirable in Khaki Campbell ducks.

The drake, when nine months old, should weigh about 5 lbs. Avoid the skinny, slightly built, Indian Runner type of Khaki Campbell, for these indicate loss of stamina and small eggs.

A great deal can be learned by handling a duck in lay. If she has a nice soft belly, which, with her posterior, is covered with a soft silky plumage, and one can place one's four fingers between the end of her breast bone and the pelvic bones (the two arches bones joined to her back, one coming either side of her vent), this shows that she is in lay, or about to lay, and is likely to be a good layer. Examine her vent, too. If this is large and moist, and the lips of the vent well developed, we have a pretty good indication that she is a good layer of large eggs. A small puckered and dry vent indicates a poor layer.

As regards measuring the distance between the end of the keel bone (breast bone) and the pelvic bones, placing one's fingers across the space is a very rough measurement, because my four fingers placed together measure 3 inches, whereas a lady's four fingers would probably not be more than $2\frac{1}{4}$-$2\frac{1}{2}$ inches. So it would be more accurate to say that a good duck in lay should measure 3 inches or over for the abdominal space.

When a duck is out of lay, the space will decrease considerably and the abdomen shrinks, so that the test is of no use. One must then rely upon fine feather texture, fineness of the bones, neck and head, and the appearance of the eyes which, besides being set high in the skull, should be prominent, so that when viewed from the front, one can clearly see the eyes protruding beyond the skull. A bird with sunken eyes or noticeable eyebrows is almost certain to be a " dud " layer.

By far the greatest number of Khaki Campbell ducks lay pure white eggs weighing round about $2\frac{1}{2}$ ozs. It is no use aiming for larger eggs than these, as birds laying very large eggs rarely lay as many as those producing smaller ones, and yet they consume the same quantity of food, and the producer rarely gets an adequate price for the over-sized eggs.

Occasionally, one gets a bird laying green eggs. This does not mean that she is not pure bred, but that she is a throw-back to her ancestors with the Rouen and Mallard blood in them. However, according to Khaki Campbell standards, the egg should be white.

The White Runner

The White Utility Runner is the White Leghorn of the duck world. It is a splendid layer of good-sized white eggs, and is an excellent winter layer. The weight of an adult bird is about $3\frac{3}{4}$ lbs., if they are really of Runner type. The illustrations shown here represent the typical Utility Runner of good type of proven laying powers.

THE BREED OF DUCK TO KEEP

Fawn and White Runners

This breed is probably the breed from which all Runners are derived, and figured largely in the make-up of the Khaki Campbell. The ducks are first-class layers, and individual birds have at times put up astounding records. The utility Fawn and White

THIS HEAD IS TYPICAL OF A PURE BRED HIGHLY PROLIFIC WHITE RUNNER

Runners of to-day are in many instances of fairly substantial build, weighing anything up to $4\frac{1}{2}$ lbs., but the average is about 4 lbs. They are hardy birds and lay a white egg. I can thoroughly recommend this breed as being, to my mind, far and away the best of the Runner breeds for the commercial egg farmer.

C

Utility and Fawn Runners

These are similar in every way to the Fawn and White variety, except that the white plumage has been bred out of them. They are not, however, taking them as a whole, such good layers as either the White or Fawn and White variety.

Standard for Buff Orpington Ducks

This breed is essentially a triple-purpose breed. It combines beauty of form and colour with good table quality and profitable egg production. It is contrary to the best interests of the breed, and at variance with the correct interpretation of this standard, to breed for any of these qualities at the expense of the other.

THE DUCK. Head: Fine and oval in shape. Skull narrow.

Eye: Brown iris with blue pupil, set high in the head, large and bold, giving the head a look of alertness and activity. A deep set, scowling, eye is objectionable.

Bill: Proportionate to the head in size. Upper mandible straight from bean to base. Colour, orange with dark bean.

Neck: Slender, of moderate length, upright.

Body: Long, broad and deep, particularly at the shoulders; free from any sign of keel. When in lay, the duck's abdomen should be nearly touching the ground. The carriage of the body should be slightly elevated at the shoulders, not quite so horizontal as the Aylesbury, but avoiding any tendency to the upright carriage of the Pekin or Runner.

The back should be perfectly straight in line, the tail being small, compact, and rising slightly from the line of the back.

Legs should be of moderate length proportionate to the body of the duck, set well apart, and bright orange-red in colour.

The colour of the duck should be a rich even shade of deep red-buff throughout, free from lacing, barring and pencilling, blue, brown or white feathers. The wings should be the same colour throughout as the rest of the body.

The weight of a matured duck in lay should be about 6 lbs.

Serious defects for which a duck *must* be passed: Any physical deformity, such as twisted wings, wry tail, humped back, twisted bill, etc. Colour other than buff, *e.g.*, white feathers in neck or breast, brown feathers, very heavy lacing, strong light line over eye, green bill.

A BUFF ORPINGTON DUCK

It is recognized to be difficult to achieve good wing colour, and therefore pale wings, though objectional and to be avoided and penalized, are not a disqualification.

A matured duck weighing under 5 lbs. or over 7 lbs., shall be ineligible for a prize at any show.

THE DRAKE. The type and general physical characteristics of the drake should be identical with those of the duck, after allowing for sexual differences. The chief sexual differences are as follows : Slightly increased length and weight, curved feathers in tail, longer bill, and colour differences, lack of depth in abdomen. The body colour in drakes should be the same as in ducks, and as level as possible throughout, with the following differences : Head and neck, seal brown with bright gloss, but complete absence of beetle green. The seal brown of the drake's neck should terminate in a sharply defined line all round the neck. The rump should be reddish brown, as free from blue as possible.

Common faults to be avoided are blue in rump, pale colour or deep brown under tail, white in wing.

Serious defects for which a drake *must* be passed are : Grey, silver, or blue head, white feathers in neck, brown secondaries beetle green on any part, very green bill, and any of the physical defects mentioned against the duck.

A matured drake weighing under 5 lbs., or over $7\frac{1}{2}$ lbs., shall be ineligible for a prize at any show.

The Buff Orpington Ducks

This breed of duck, if bred true to type, is undoubtedly the handsomest of the utility varieties. Although not such a prolific layer as either Khaki Campbells, White, or Fawn and White Runners, nevertheless, it is a true dual-purpose bird, since it attains a very useful weight, and compares very favourably as a table bird with other breeds bred entirely for table purposes.

This breed of duck was originated by that remarkable breeder of poultry, William Cook of Orpington, some time in the eighties, who was likewise the originator of the Orpington fowls. The Buff Orpington duck is one of the few breeds that still adheres to the original type as introduced by its originator.

THE BREED OF DUCK TO KEEP

The majority of the ducks lay a large white egg in these days, although originally they laid mostly a green-tinted egg, but owing to the consuming public preferring white eggs, breeders have gradually bred out the green egg factor.

The standard as laid down by the Buff Orpington Club is given above.

The aforegoing standard describes very clearly and fully the salient and necessary points at which to aim when breeding this duck, so that further comment is unnecessary. However, it is as well to point out to novices that they must not expect to hatch out ducklings that are all perfectly marked should they purchase a setting of eggs, even if they come from very well bred stock.

There is always a percentage of rather light-coloured ones, and some that may revert back to a certain extent to the wild duck type.

None but well-marked and true-to-type birds should be used as breeders, as otherwise the faults will be continued and exaggerated in the resulting progeny.

The Magpie Ducks

This breed at one time bid fair to become a very popular one for a while, and still is amongst poultry fanciers, for it requires skill to produce birds that are accurately marked.

As the name implies, these ducks are pied, that is, black and white. They are moderately good layers, and run up to 7 lbs. in weight.

The Club standard for this breed is as follows :—

Bill : Pale yellow to deep orange. The neck white ; the head likewise, except for a black cap covering the whole of the crown of the head, but clear of the eyes.

Breast : White. Back and tail black.

Legs : Yellow.

Eyes : Dark grey. The illustration shows the markings better than words can describe them.

There are a few other breeds of ducks that have been recognized by the Poultry Club, such as Abcot Rangers, Coalvy Fawns, but they are not worth considering, being nothing more than a Khaki Campbell sport in the case of the former, and Coalby Fawns likewise, or with some Buff Orpington blood introduced, but the latter so closely resemble many pure bred Khaki Campbells, that it would take an expert to distinguish them.

A WELL-MARKED MAGPIE DRAKE

Ducks for Table Purposes Only

A few years ago, I should without hesitation have said that in this country the Aylesbury duck easily came first as the most popular table duck in the trade. But conditions and customs have changed much during the last ten years. Families and houses are becoming smaller, and persons are purchasing more and more pre-cooked foods which just require heating up to be ready for the table. The result is that there appears to be less demand now for the large, high quality, white table duckling, than existed only a few years ago.

It may be thought that the purveyors of cooked meats will find that it pays them best to use large ducklings, which will be cooked and sold in portions at so much a pound, than to use small ones, which cannot conveniently be divided up into so many portions.

For the time being, at any rate, if one looks down the daily Smithfield market reports, which are published in several of the principal daily newspapers, one will see under ducks: " Aylesbury ducks " quoted at so much, " Country ducklings " at another price, the former generally being quoted at a rather lower rate. This is because in the trade, ducks other than large white-feathered and pink-billed ones are regarded as " Country ducks," or birds of rather lower quality. Further, the trade still regards the large white duck of 5 lbs. and over as being an Aylesbury.

I have already discussed this point of the apparently changing demand for a smaller duck towards the end of Chapter I, and whatever the change may lead to, I am convinced that the pure Aylesbury duck will still be used as the basis from which smaller white ducks will be bred. My reason for believing this is because there is no other breed of duck in this country, or any other country for that matter, in which the ducklings will attain such size and so much flesh in so short a while. This is a most important factor in breeding, be it for beef, pork, mutton, or poultry for the table. This propensity to rapidly turn crude foodstuffs into meat of high quality, is one which breeders in this country have aimed at and excelled at for generations. The Aylesbury duck, in my experience, is paramount as a flesh maker, compared to any other breed of duck, so that we cannot do better than continue to breed them, either to be used for crossing purposes where a smaller bird is required, or for marketing as a pure Aylesbury where birds of a large size are required.

The Aylesbury Duck

Strictly speaking, there are two varieties, the pure utility type and the exhibition type. There is in practice a vast difference between the two. For commercial purposes, duckling rearers confine themselves to the former. It is a more active bird, far easier to breed and fatten rapidly than is the immense exhibition type, with its tremendous depth of body and long keel. The illustrations clearly show the difference in type.

The only use of the exhibition type to the commercial breeder, is occasionally to introduce some exhibition blood into his commercial stock in order to maintain size. Otherwise the show variety is not a commercial proposition, since it possesses an enormous frame which takes a long while to cover with flesh, and consumes a great deal of food in doing so. This is not to say that the birds will not attain a very much greater weight at a given age than will the commercial variety, but there is too much bone and not sufficient flesh about them as youngsters.

I think it would be no exaggeration to say that if one wished to get as much flesh on an exhibition type of Aylesbury, as one can on a commercial type of bird at ten weeks, one would have to run the former on to fourteen weeks, and feed to it at least 25 lbs. more food, and food of a most nourishing kind, such as sausage meat, which is the food given to this large variety when being prepared for exhibition. I was once talking to a successful exhibitor at the Dairy Show about one of his ducks, which was shown in the dead table poultry class. The carcase weighed over 12 lbs. plucked, and had won first prize. He informed me that he reckoned that this particular duck had cost him close on £5, taking into consideration the cost of feeding the parents, and the fact that he had

had to give it pounds of sausage meat to eat in order to get it up to this immense size!

A useful type of commercial Aylesbury duckling should weight 5 lbs. when killed at the end of the ninth week, that is, when plucked and completely cold. The drake will weigh about 1 lb. more. First quality breeding stock should produce ducklings of rather higher weight than this, $5\frac{1}{2}$ lbs. for ducks, $6\frac{1}{2}$ lbs. for the drakes, that is, assuming all has gone well with the rearing, feeding and management. As adults, really good commercial stock should weigh at the commencement of the breeding season 7 lbs. in the case of the ducks, and 9 lbs. the drakes, that is, when they are about eight months old. Second season birds may exceed these weights. At the end of the breeding season, there will be a reduction in their weight, anything from 1 to 2 lbs.

In the case of exhibition Aylesburys, when a year old they will weigh anything up to 15 lbs. in the case of the drakes, and 11 lbs. for the ducks, but that is for first-class stock. 11 lbs. and 9 lbs. for drake and duck respectively, are quite usual weights.

Taking them as a whole, Aylesburys are not grand layers, and cannot compete with the lighter breeds in this respect. Further, they do not, as a rule, lay all the year round. For instance, if hatched in February, the ducks will probably commence to lay in July, and produce a dozen to twenty-four eggs, and then fall into a moult, when no more eggs will be forthcoming until the end of December.

If ducks intended for stock purposes are hatched in May, they generally commence to lay in December, although this may vary according to strain, locality, weather conditions, and method of feeding.

Occasionally one hears of Aylesburys producing over 200 eggs a year, but birds that do this are rarely pure bred stock, but have at some time or other been crossed with Runner, White Campbell or Pekins, but if this cross took place many generations before, it is often difficult to discern any trace of alien blood, except that one may get unevenness in the size of the progeny and a tendency to yellow bills.

Fifteen years ago I trap-nested two pens of commercial Aylesburys and two pens of exhibition type. Amongst the former I had one bird that laid 187 eggs in the first year, but the average was 118. The exhibition pens showed a lower average, namely, 94 eggs each, and one bird only laid 43, but all the eggs were very large—5-6 ozs. each.

On the whole, I should consider an average of 100 eggs per bird satisfactory, providing fertility was good.

My experience of trap-nesting Aylesburys for a few seasons, was that selecting for higher egg production had a tendency to produce a more active type of bird, whose progeny did not fatten up so rapidly as did those from more stolid and less temperamental birds. It seems far better, if one is going to trap-nest, to aim at selecting ducks that show a high percentage of fertile eggs and good hatchability from them, than to produce those that lay large quantities of eggs, as my experience rather pointed to less satisfactory results being obtained from these latter as breeders.

Although a considerable amount of experimental work has been done by breeders of Aylesburys in order to improve laying qualities, coupled with good fertility and hatchability, the results have not been altogether consistent, and failure to secure the desired results are not easy to explain. For instance, when trap-nesting, one may find that certain ducks are producing eggs which give very much better hatching results than one is obtaining from others, but then, for a while, the other batch suddenly commence to give a series of eggs producing excellent results, and may continue to do so for some while. This makes it very difficult to lay down any hard and fast rule in respect to these matters.

From what I have been able to observe in connection with varying fertility and hatchability from certain ducks, it seems to me that quite often these sudden

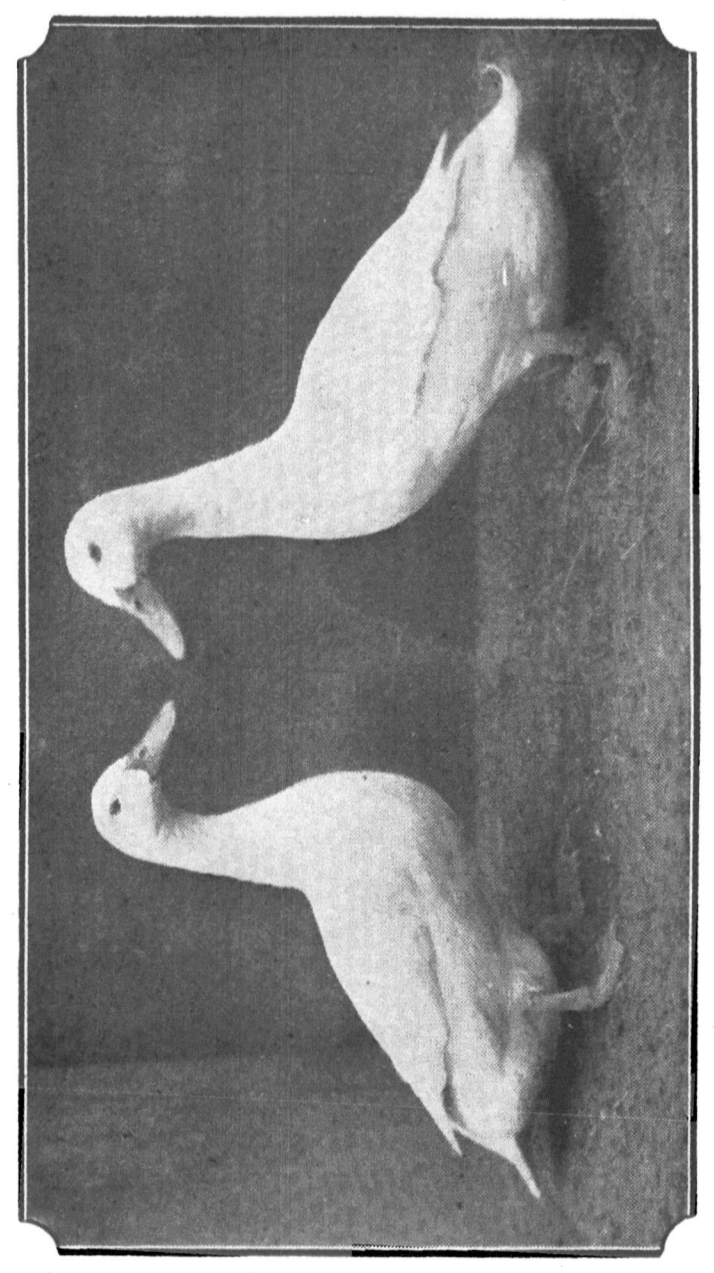

A GOOD TYPE OF UTILITY AYLESBURY DUCK AND DRAKE

FINE UTILITY AYLESBURYS

THE BREED OF DUCK TO KEEP

variations are due to the fact that the drakes take a particular fancy for certain ducks, and tend to neglect the others. Those ducks that become favourites go through a rough time of it and lose condition, with consequently bad results in the fertility of their eggs. The neglected ducks also may lay a series of infertile eggs. When I have noticed such an occurrence, I have removed, temporarily, one or two favourite ducks from the flock, which has had a beneficial result on fertility, for the drakes have then turned their attention to a number of other ducks instead of a few.

Then again, local adverse conditions may produce poor fertility. Access to too much cold water, exposure to cold winds, lack of undisturbed rest, as a result of being frightened by visits from foxes, owls, cats, or a considerable number of rats.

Another source of poor fertility is when drakes or ducks get large warts on their feet. Apart from the mere presence of the warts, the pain causes loss of condition in both sexes.

These are only a few possible causes and reasons for poor fertility and consequent poor hatches. At any rate, I am convinced that lack of physical condition on the part of either the ducks or drakes, or possibly both, is generally the cause of poor hatching results, rather than any inherited disability or natural tendency to barrenness. Again, a duck that has laid many eggs cannot always provide the contents of these eggs with all the essential elements that a germ requires for its development when it commences to grow.

In the case of small pens of ducks mated to one or two drakes, I have had cases where one drake has ruptured its male organ, by no means an uncommon occurrence, and the other male has been sterile, resulting in complete infertility of all the eggs from that pen.

Aylesbury ducks do best when kept in a nice sheltered orchard, upon well-drained ground, and they prefer warmth to cold. Further fertility from ducks kept without access to swimming water is likely to be as good as, if not better than, those that have access to deep ponds or streams; especially is this so during cold weather.

Whilst on the subject of fertility of hatching eggs, it is as well to remember that it is best to mate up one's pens in October or November, so that males and females are thoroughly at home with one another before eggs are required for hatching purposes. Ducks that have been running without drakes, or perhaps with one only, will resent the introduction of strange drakes. It is a great mistake, too, to introduce a young drake into a pen where there already exists an older drake. I have known both ducks and drake to bully the newcomer, and even to drown him when there had been a pond handy. Such goings on in a pen of breeders is not conducive to fertility.

A friend of mine, who has had over half a century of experience as a professional Aylesbury duck breeder, makes a point of mating his flock up early in autumn, and since he has remarkably good results, I followed his advice and did likewise, for experience in matters of poultry breeding and rearing counts far more than anything else.

If breeding ducks are to be kept in good condition, they should have somewhere in which they can wash themselves. It may sound fantastic, but there is little doubt that a white plumaged breed of duck becomes thoroughly miserable if it cannot keep itself clean and white, and as ducks are very sensitive creatures, it is more than likely that feeling dirty and looking dirty upsets them, with detrimental results all round.

When the ducks are being kept in ordinary grass runs, it is easy to construct a small concrete-lined pool or to purchase, which one can for a few shillings, an old dripping-pan type of bath, and sink this in the ground. It does not matter if the bath has a few holes in it when purchased, as these can easily be filled up with cement or pitch when it is let into the ground.

THE BREED OF DUCK TO KEEP

It is more satisfactory, though, to make the pool in a raised pile of soil. In the bottom of it make a drain-away with a plug, such as one finds in old-fashioned baths. A pipe can easily be connected with this, so that the water can be drained out as and when required. It is not healthy for ducks, especially young ones, to drink a lot of filthy water during warm weather, and the water soon does get very dirty and contaminated with the ducks' droppings, when many ducks have access to so small a pool.

As regards mating up one's breeding ducks, my own preference is to use second season ducks and well matured first season drakes, although healthy second season drakes are often quite satisfactory; but they are not always so active during the winter months as an eight months old drake.

During the early part of the breeding season, that is, between December and the following February, mate three ducks to a drake if small numbers only are being run together, *i.e.*, four drakes to a dozen ducks. If large breeding pens are being used, then five drakes to twenty ducks will probably give quite satisfactory results. In practice, I have found that excellent fertility can be obtained with twelve drakes mated to fifty ducks. For commercial purposes, the latter-sized breeding flock is as convenient as any, although I know of commercial breeders, who run them up to double this size with success.

About the month of April, the drakes will become very active, and it is often good policy to reduce the number of drakes in the large pens to one drake to five ducks, otherwise one may experience poor fertility as a result of the ducks being damaged and worried by over vigorous males. I don't advise getting rid of these surplus drakes, but consider it best to keep them in a separate pen right away from the ducks. If any drakes amongst the flock begin to look out of condition, take them out and rest them, replacing them with the spares; the latter will not be strangers or strange to the runs, therefore they will commence mating almost at once with the ducks. Ducks, as a race, have long memories.

As a rule, one cannot count on getting fertile eggs from the first few eggs laid by ducks in a mated pen, even if the birds have been running together for some while. If the ducks have been unmated, and drakes are put in with them, if may be ten days before the eggs are fertilized, for, as I have already mentioned, it takes some while for the birds to become accustomed to one another, so have patience if you find a large percentage of infertile eggs to commence with.

All eggs should be collected as early as possible from the houses of the breeding pens during frosty weather, as it does them no good to be in the cold for hours, since ducks lay very early in the morning—at least, most of them do. I have seen ducks lay eggs at 4 a.m., and comparatively few eggs are laid after 8 a.m.

All hatching eggs should be carefully stored, as careless storing may seriously affect hatching results. Poor hatching results can frequently be traced to the fact that the eggs have been subjected to too cold, too hot, or too dry an atmosphere prior to being put under hens or in incubators. The best temperature for storing them is one of about 50° Fahrenheit or 12° Centigrade. Unless specially stored, hatching results will deteriorate rapidly after ten days; this is generally due to loss of the moisture contents in the eggs. For best results every egg should be wiped clean with a damp rag dipped in warm water. Then inspect the egg carefully for any minute cracks, which are often very difficult to see unless the egg is quite clean. If a crack is noticed, smear the crack with a celluloid solution such as Durafix. This will seal it; otherwise, if this is not done, the egg will never hatch out. When the eggs are clean and dry, wrap each one in greaseproof paper or waxed paper, such as one purchases for wrapping apples in. This paper is inexpensive, and can be used over and over again. Having done this, lay them on their sides in a box upon a layer of bran, and cover them with a layer of hay. Every day turn the box over. This will cause the yolks to move into a different position, and so prevent any chance of their sticking to the inner membrane of the shell, for if this happens they will not hatch.

Providing the eggs are strongly fertilized and are stored as described, I know that it is possible to get quite good hatching results from them, even after they have been kept a month, for I have exported hatching eggs to South Africa, which have arrived safely and hatched out with over 50 per cent results.

Exhibition Type of Aylesbury

I have already made some remarks regarding the difference between exhibition and utility types of this breed.

For exhibition purposes, an Aylesbury drake should weigh at least 10 lbs. when six months old, and the duck 8 lbs. These weights, though, would be insufficient to win with at anything but the smaller shows. In shape the sexes are similar. The only noticeable difference is the somewhat stouter formation of the head and neck, and the fact that the drake carries four curled feathers above his horizontal tail feathers. In both sexes the head should be long and straight, the bill pink, long and broad and not dished.

The body should be massive with good girth, and with a deep straight keel extending from the breast to just behind the legs, so that when the bird is on the alert and viewed sideways, the line of its neck should be almost at right angles with the tail.

The chief defects when showing are a yellow bill, brassiness in the plumage, that is, a yellowish tinge, and too upright a carriage. All these faults are likely to be due to alien blood, although pure Aylesburys are inclined to go yellow in the bill and get a yellow tinge in the feathers if allowed free grass range, and are given maize to eat and have no swimming water. Second season birds frequently get black specks on their bills, known as peppering. This is a defect.

DUCKS AND DUCK BREEDING

If one intends to show any birds, these should be selected early and kept off grass, penned in a shady spot and fed on white food, such as middlings, barley meal and meat scraps. Have a fairly deep trough handy filled with gravel, sharp sand and water. This helps to keep the bills pink, since continual sozzling in the sand rubs off any yellow skin that may form upon the bill. Incidentally, although not strictly permissible,

A PRIZE-WINNING EXHIBITION TYPE AYLESBURY DRAKE. NOTE THE DEEP BUST AND LONG KEEL

before shows, the bills are treated by rubbing them with finely powdered pumice stone and damp rag and finishing off with an oily rag. Furthermore, the birds are allowed to have frequent baths in absolutely clean soft water. This means having to fix up a little sunken bath with a cement surround to prevent dirt getting in. The birds should be absolutely dry before being put into their show baskets, for, if damp, dust will adhere to the feathers and spoil any chance of winning at a large show where competition is keen.

The usual procedure, when selecting birds for show purposes, is to pick out two or three of the best a fortnight or so before the show. Keep them in show pens all night and part of the day, in order to train them to become used to being approached by persons. Feed them very well, and add a little cod-liver oil to their food, as this helps to increase the whiteness of their plumage. Then, about four days before the show, start letting them have their extra washes in frequently changed water. After finally selecting the birds you intend to send, make sure that their feet and shanks are spotlessly clean, likewise their bills, and give the latter a last wipe over. Condition in which they are shown counts tremendously in the eyes of the judges. A duck previously trained, in a show pen enables the judges to view the bird to its best advantage. If it is untrained and frightened, it will not stand still when the judge inserts his judging stick into the cage and touches its plumage, and he may get a wrong impression of its quality and knock off marks, so that the little extra trouble in training the birds is well worth while.

Any Other Variety Duck

Descriptions of the principal utility breeds of ducks only have been given, but, of course, there are many other varieties. Some of them, such as the Cayuga, make an excellent table bird, the flesh being exceedingly tasty, but they are not a commercial proposition, since they are of little value if killed before attaining an age of sixteen to twenty weeks, when they reach a weight of about 6 lbs.

The Cayuga

The Cayuga, like the East Indian duck, is essentially an ornamental water fowl. Both are exactly alike as regards plumage, having lustrous beetle-green-black feathers. The East Indian duck is, however, very much smaller, and is a delightfully dainty and well proportioned little creature. I have a very soft spot in my heart for the East Indian, for as a very small boy I kept this breed. They were very wild and did not make friends easily, but on the ornamental pond, with the sun shining upon them, they looked almost fairy-like, so iridescent is the sheen of their plumage.

The Mandarin

However, to my mind, the gem of all the water fowl is the Mandarin. The beauty of this little bird defies description, so full of colour is it. Kingfishers and humming birds look drab against it.

The drake has a large long crest pointing back, like a peewit, which it can raise at will. This crest is green and purple, shading into a chestnut colour towards the top. A wide stripe traverses the head above the eyes, like great eyebrows. The neck has a ruffle of almost red feathers, and the side of the neck and breast are a beautiful claret hue. The wings at the shoulder joints have a curious fan-like shield which stands erect, coloured bright brown tipped with green or blue. The bill is crimson and the legs pink. The duck is of rather plainer plumage.

These little water fowl are the Love Birds of Duckdom. When mated the pairs show most remarkable signs of affection, and, like Love Birds, they rub and stroke one another's heads and bills. In China they are presented to bridal pairs, as an emblem of conjugal affection. However, they are great fighters, and once they are paired off no attempt must be made to separate them and mate one or the other to another bird, for this will certainly result in the death of a female, who will be killed by her strange mate.

They are quite hardy, and can be kept in the open all the year round, if given a little shelter in winter. They will eat almost any kind of grain, wheat and buckwheat.

The Pekin

As a layer the Pekin is far better, taking them as a whole, than Aylesburys, but in spite of this and being a good table bird, as well, they are not popular in this country. In the U.S.A., on the other hand, they are by far and away the most popular table bird, being bred in tens of thousands on some of the large duck ranches. The chief reason for their failure to be popular in England, is undoubtedly due to the poulterer's prejudice to a yellow-skinned bird, and also because they rarely pluck as cleanly as do Aylesburys, which gives the dressed duck an untidy and rough look when displayed in the poulterer's shop. It is said that they are as quick growers as Aylesburys, and will attain the same weight in the same time as these, but an Aylesbury duck raiser who kept some tells me that they make a very poor show if killed at ten weeks and are packed in the same hampers as Aylesburys of the same age.

To look at, the Pekin in some respects resembles an Aylesbury with rather an upright carriage, but the plumage is totally different, being of a distinctly yellow colour. The bill, instead of being long, straight and pink, as in the case of Aylesburys, is a deep yellow, and is decidedly shorter and curved, but not sufficiently to call it " dished," that is to say, curves down to the centre and then rises again slightly.

The head is large and much rounder than that of the Aylesbury; the latter has a flat top to its head, whereas the Pekin is rounded, in fact, is rather high at the top of its pate.

The eyes are rather deep set, and the neck is very massive and of good length, but, owing to its thickness, does not appear to be nearly as long as that of the Aylesbury.

The breast is wide and deep. There is no keel to the bird worth calling one, and the belly, owing to the upright carriage of the bird, comes very close to the ground.

One of the characteristics of the Pekin is that it has very puffy cheeks, appearing almost to be swollen, and this peculiarity is one that often crops up for generations in Aylesburys, where crossing with Pekins has taken place years before.

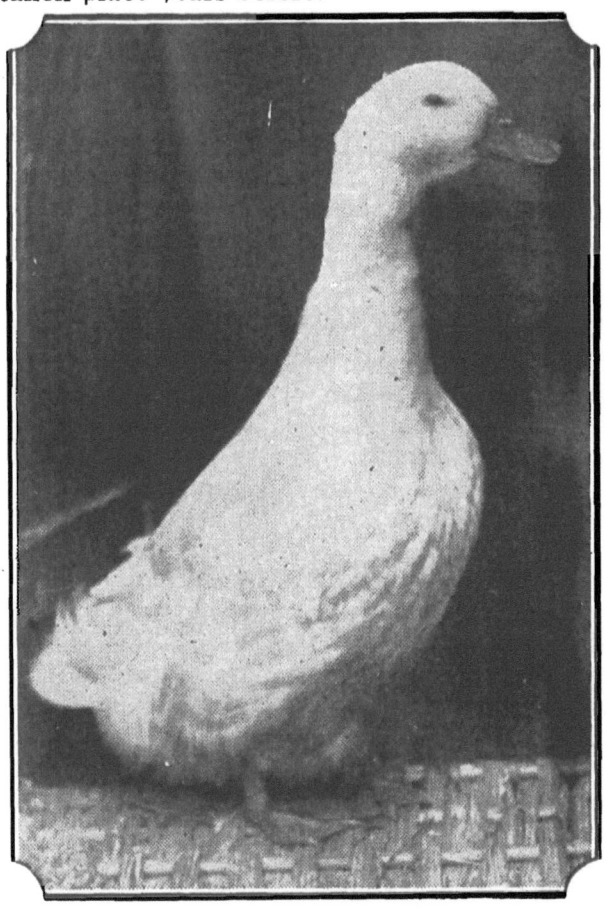

TYPICAL PEKIN DRAKE. NOTE THE PUFFY HEAD AND LOOSE FEATHERING

The Pekin is a very profusely feathered bird, and carries rather a wedge of feathers over the thigh.

As far as weights go, for show purposes the drake should scale about 9 lbs. and the duck 8 lbs.

THE BREED OF DUCK TO KEEP

The Pekin has one great merit, and that is hardiness. The eggs, as a rule, prove very fertile, easy to hatch, and the ducklings rear easily. For mating, it is usual to allow four to five ducks to a first year drake.

In order to get a good comparison of the respective merits of Aylesburys and Pekins as table birds, one should visit the table poultry section at the Dairy Show, where there is always a wonderful display of dressed but not drawn ducks. One can pick out the Pekins at once from the Aylesburys by the very yellow colour of the flesh and creamy neck feathers, and as a rule, at this Show, the specimens of Pekins compare very unfavourably with the Aylesburys for size and quality.

The Rouen

Undoubtedly the Rouen is by far and away the most beautiful of the utility ducks, in fact, anyone requiring ornamental water fowl for a pond might very well choose Rouens as not only being capable of fulfilling the role of an ornament, but also as being table birds second to none. Personally, I prefer the looks and the flesh of the Rouen to any other breed, but, like all good things, they have their disadvantages from the commercial man's point of view, who has to derive his living from the sale of eggs or table poultry. Their chief disadvantage is that they are, taking them as a whole, poor layers and rather late layers. That is to say, it is difficult to get them to lay freely before February, which, coupled with the fact that they are not such quick flesh-makers as either Aylesburys or Pekins, puts them out of the running for the early duckling trade. Then, again, not being white feathered, they come up against the prejudice of the trade, who dislike dark plumaged birds, or birds with a dark skin.

These disadvantages are, of course, serious ones in the eyes of those who are rearing ducklings commercially. But for private persons who like to see beautiful things around them, and at the same time require ducks for table, they cannot do better than keep Rouens.

The drakes of the exhibition type Rouen weigh up to 11 lbs., and a most splendid fellow he is to look at, with his beautiful metallic green head, claret coloured breast, and a white ring round his neck and body colour a soft French grey. A clearly defined rich blue band crosses his wings, with a snowy white thin bar either side of the blue. He has bright brick-red legs

AN EXHIBITION TYPE ROUEN DRAKE

and feet, bill of a yellowy-green colour, with a black bean at the tip. These are the points most obvious to the casual onlooker, but for the fancier who loves beautiful details, it is the fine and even chain armour pencilling in black across the delicate French blue body colours, free from all white or rustiness and other details, that are of interest, but which would not appear to be of importance to the ordinary observer.

The duck generally weighs from 7 to 9 lbs. Her plumage is very sombre compared to that of her mate.

The bill is a bright orange with a black bean at the tip. The head is a rich brown, the eyes hazel. A wide black line runs from the base of the bill to the neck, and running along either side of the head, above and below the eyes, is a bold black line. The body feathers are also a rich brown, each feather being distinctly pencilled, and on the wings is a blue band similar to that of the drake; the legs are an orange brown. There is no white ring on the neck of the duck, and to have one is a defect.

For breeding, it is not advisable to give a drake more than three ducks. The large eggs are generally green.

The Muscovy Duck

The Muscovy is the only other really good table duck, of what one might call a non-composite breed, that is to say, of a breed that has not been formed by crossing the various breeds I have mentioned. As a matter of fact, the Muscovy is more of a goose than a duck. Further, it has the peculiarity of liking to roost in trees, where it will nest if it gets the opportunity. Another curious characteristic about this breed is that the eggs take nearly a week longer to hatch than do ducks' eggs.

The male birds can be made to attain a great weight if kept up to six months, 10 to 12 lbs. not being at all uncommon.

The birds have a peculiar smell of musk about them, but this is not noticeable when the flesh is cooked.

The Bali Duck

A new variety of duck has lately been introduced into this country. It is called the Bali duck, and was imported by Misses Davidson and Chisholm from Malaya in 1925. The ducks derive their name from an island off the coast of Java. To look at, the duck resembles a White Runner duck, with a somewhat dished bill, but the peculiarity about them is that they have a topnot of feathers right at the back of their head, giving them a most quaint appearance. They are good layers of white eggs.

HOW TO MAKE A START ON A SMALL SCALE

Of all the questions asked me in the many letters I have received from correspondents, the query which has been most frequently put to me is, " How can I make a start on a small scale ? " Therefore, I intend to deal with the subject as fully as space will allow, and divide the chapter into three sections. The first section will be for those who only wish to keep ducks on quite a small scale. Some of the ground relating to the subject has already been covered in Chapter I.

If the intention is to keep about half a dozen laying ducks to supply a small household with eggs, and have an occasional duck for use in the house, then keep either Khaki Campbells, White Campbells, or Buff Orpington ducks, preferably the former.

Buy Day-olds

The cheapest way of starting is to purchase a dozen day-old ducklings, if you have no broody hen available, otherwise a hatching of eggs might save you a few shillings; but I would rather buy half a dozen live ducklings than pay for a setting of twelve eggs. By the time the latter have been banged about in the post, or, if they have been sent by rail, then banged and thrown about by careless porters, one is lucky to hatch out any ducklings at all ! Further, if one buys day-old ducklings, and they arrive safely and in a healthy state, that finishes the transaction, whereas purchasing a hatching of eggs is only the commencement. Some of the eggs may be broken or cracked slightly. This means that a claim has to be made, letters written, time wasted in getting replacements for the broken eggs, also for any infertile ones later on. Therefore, buy day-olds every time. If possible, get the ducklings from a really reliable breeder who can sex the ducklings, and if you can afford to do so, only purchase " ducklets," the " drakelets " are just a nuisance. You will have to pay exactly double the

HOW TO MAKE A START ON A SMALL SCALE

price if you purchase all " ducklets," because the sexes average out fifty-fifty; the breeder has to spend time on sexing the ducklings, and you must pay him something for his skill. He doesn't want the drakelets any more than you do. I have sexed thousands of ducklings for customers, and know something about the difficulties of disposing of the drakelets. I used either to kill them off, if they were hatched on an inconvenient day of the week, or else, if hatched about a day or so before the local market day, I sent them in there and disposed of them at the rate of about 8s. per 100, not even the price of the eggs before they were hatched.

Purchase Good Stock

At any rate, do take my advice and purchase really good stuff. Don't spoil the ship for a ha'porth-o'-tar. Rather purchase half a dozen ducklings from a really good breeder, than a dozen from an unreliable source. If you are not sure where to get them from, write as a last resort to me at Turner's Wood, Chalfont St. Giles, Bucks., and please enclose a stamped and addressed envelope. Incidentally, I cannot be certain of helping readers with addresses of likely sources of supply, for the demand for day-old ducklings, young ducks and stock birds has been tremendous since the war. I alone have had over 600 enquiries since this book appeared. Unfortunately, owing to my war work, I am unable now to produce ducklings or stock birds on a commercial scale. Until the end of the war I cannot be regarded as a source of supply.

In these times, prices for really good ducklings are not very reasonable. You should be able to get sexed ducklings, with sex absolutely guaranteed, for 5s. each. With care, you should be able to rear every one, so that for the very moderate outlay of about 30s. you can have half a dozen ducklets from well-bred parents. For the outlay of another £1, these can be reared up to laying stage. But in these days people are becoming dreadfully lazy, and require everything served up and ready to eat, so to speak. Therefore, if you cannot be bothered to rear day-old ducklings, your next choice is to buy young pullet ducks at about eight weeks old. These, from quite good stock, cost about 8s. each pre-war, but double that now in 1941. You will certainly see them advertised at much cheaper rates than this, in an advertisement such as this: " For sale, 20 lovely Khaki Campbell ducks on the point of laying, bred from Roscoe's famous 317 egg strain ducks. Sacrifice 4s. each. So-and-So Duck Farm." Heaps of people will be caught by such an advertisement, especially if I had an advertisement in next to this one, and I was asking 10s. each for guaranteed pullet ducklings bred from trap-nested parents with egg records over 200 eggs.

Use Deposit Systems

The novice would say: " Here's Roscoe asking twice as much as So-and-So is for Khaki Campbell ducks, and what is more So-and-So's are Roscoe's own strain, only from better class birds. Not likely, I'm not going to pay 10s. for ducks just to say that I got them from Roscoe, when I can get the same, if not better, from another man for 4s." I know that this has happened many a time, because people have written and told me that this has occurred to them, and on one occasion a clergyman came round to my place and said he wanted to purchase some young Khaki Campbell ducks. He had come personally, because he had ordered some Khaki Campbell pullet ducks, said to be my strain, and when these ducks had arrived he noticed that (*a*) they came from Liverpool, whereas the man from whom he had ordered them had a duck farm at the opposite side of England ; (*b*) that the birds looked old ones and were moulting. It was September, and the ducks were supposed to be five months old and on the point of laying. He became very suspicious, and wrote to the address in Liverpool from whence they had been sent. He received a reply

HOW TO MAKE A START ON A SMALL SCALE

from a market poultry salesman, to say that all he knew about them was that he had had a commission from a client to buy up any Khaki Campbell ducks being sold in the market, and to hold them for him until he got orders as to where to dispatch them. The clergyman then took the trouble to motor over to the so-called "Duck Farm." When he got to the village where it was supposed to be, no one had ever heard of it. So he went to the post office, and they told him there that there wasn't actually such a place, but that a man at one of the cottages carried on a poultry dealer's business and advertised ducks and chickens for sale, but didn't actually keep any.

I could tell you of numerous other similar cases of fraud that correspondents have written and told me about, so that I do strongly advise purchasing only from a reliable source. Further, if the purchases are made as a result of an advertisement in one of the poultry journals, be wise and take advantage of the deposit system which most of them offer as part of their service and protection to their readers, for, as I have already shown, there is plenty of fraud going on. Beware of very cheap offers of Khaki Campbell ducklings, as these in all probability have been sexed, and you would find yourself landed with 90 to 100 per cent drakes.

Do Not Over-pay

Having advised you not to purchase very cheap stock because you are likely to get rubbish, I am going to tell you, likewise, not to be foolish and pay very high prices for utility birds from people who undoubtedly have good stock, but open their mouths far too wide. I have seen a breeder, who had only won fifth place at a Laying Test, ask £3 3s. per dozen for day-old ducklings, whereas the man who won the Test was only asking £1 1s. per dozen, and this man was definitely a better and more reliable breeder than the former. So don't think that you are necessarily going to get something really super just because a man is asking a fictitious price for his ducklings.

Popular Prices Best

If later on you decide to sell hatching eggs, day-old ducklings and stock birds, you might bear this in mind, namely, that it pays you much better to sell the whole of your output at a reasonable price, than it does to only sell a quarter, or probably not even that, at double the price. You limit your market very much if you charge very high prices, even though you may have been successful at laying trials and exhibitions.

As a proof of this, I sold over 10,000 hatching eggs from a flock of good, well-mated Khaki Campbells, in one season at the rate of 5s. per dozen. Against this, from a few pens of good trap-nested Khaki Campbells with high records, I charged from 15s. per dozen for their eggs, according to their pen records, and only sold sittings to the value of £11 in a season. As a matter of fact, I did not waste any of the eggs not sold from these pens, as they were hatched and the ducklings used later on to produce more stock for my cheap sittings. I know that my experience is not an unusual one, because a man who had been a most successful competitor at Laying Tests, told me that he had been asking £1 per dozen for hatching eggs from his best pens, and had only sold £5 worth in a season. I am convinced that had he charged 10s. per dozen, he would have sold twenty times as many, for his stock was excellent.

So when you eventually blossom out as a breeder and vendor of hatching eggs, etc., aim at selling good stuff at a popular price. Rather than be greedy and hope for 500 per cent profit, be content with 50 per cent—that is still a big one.

It Pays to Advertise

And don't forget to advertise your wares. You will sell very little unless you do. Furthermore, when you have plenty of produce for sale, advertise regularly, so that your name becomes familiar to readers. Do not expect to sell all you have through one insertion. Try to be original. Even if you are

HOW TO MAKE A START ON A SMALL SCALE

only advertising in the " smalls," it pays to spend an extra shilling or two in order to attract attention to your particular advertisement. For instance, if you are advertising Khaki Campbell hatching eggs for sale, the heading *K.C. Hatching Eggs* is not attractive, but *Rain ! Rain ! Rain !* attracts the eye at once. You can then proceed, " It's never going to leave off raining, so you had better keep ducks. My stock birds are weatherproof and lay all winter, wet or fine, etc." I inserted a " small " worded in this way one spring, and sold hundreds of hatching eggs as a result, and customers mentioned that the advertisement had attracted them. My experience proved to me that attractive and full advertisements, inserted in the classified advertisements columns of poultry journals, gave me better results than display advertising on a small scale.

If you decide to buy day-old ducklings, whether they be Runners, Khaki Campbells, or any other light or medium weight ducks for egg production purposes, buy them between mid-April and May. If you buy hatching eggs, then purchase them a month earlier than this. It is always advisable to order day-olds well in advance, because the demand is great during April and May, and you might have trouble in getting delivery of them.

Make Sure of Broodies

Before ordering hatching eggs, make sure that you have a broody hen available to put them under when they arrive, and make certain that she is good and broody. Keep her sitting on some nest eggs, or even peeled potatoes, for two or three days. If she is a large hen, you can safely put a dozen Khaki Campbell eggs under her. A small hen should not have more than ten to sit on.

If you have day-olds arriving, and you intend to put them under a broody hen, she, too, must be very broody, and have been sitting on nest eggs for several days. The ducklings should be put under her at night, otherwise there is always the danger of the hen killing them. I have known hens become frightened of suddenly finding some curious dark squeaky objects put under her in the daylight. Once a hen has brooded some ducklings overnight, she soon realizes that they are harmless and will continue to nurse them.

Emergency Brooder

Ducklings, being very hardy, can be kept without a broody hen. A hot water bottle, put into a wooden box with plenty of hay, will keep warm for a long time, and the ducklings will obtain all the warmth they require from this, until something more satisfactory can be arranged for them.

Broody hens can generally be purchased from local poultry farms for about 5s. each in peace-time, if one's own are not to be relied upon.

Book Stock Early

It is not advisable to purchase young ducks later than the end of August, unless one is quite certain of the dependability of the person from whom the purchase is being made. The reason for this is that many duck-keepers clear out old birds at the end of the season, namely, July, and these old birds are bought up by dealers and kept until through the moult, and then sold again as young ducks. It is not easy, even for an expert, to say definitely how old a duck is once it has completed moulting. If, however, one buys April hatched ducks in August, they will have completed their first moult, which takes place when they are ten weeks old, and will have an unmistakable air of freshness about them, which an old duck never has when it is about to moult.

Now, having once bought your four or five young ducks, you only have to feed and house them properly, and by mid-September you should be getting eggs, and by October they should be laying with clocklike regularity until the following July, when they will moult. But these birds will be good for another two seasons at least. For some reason or other, when kept in quite small numbers, second and third season ducks will often lay 200 eggs and over, but this rarely happens if one keeps flocks of them.

It will be seen that if one keeps a few good ducks for three seasons, and during that period they average 600 eggs each, as they easily can do, they will certainly not owe one anything, and whatever one gets for them at the end of the third season will be profit. Furthermore, they will have shown a very handsome profit over cost of feeding, after deducting the whole of their original cost.

On More Ambitious Lines

If it is intended to keep ducks in considerable numbers, I would always advocate a person buying two, three, or half a dozen breeding pens, and working up a stock from these. If the second or third season ducks were bought in August while moulting, they can be purchased at a very moderate price from a reliable breeder, for probably 7s. to 10s. each in peace-time. For Khaki Campbells, one drake to six ducks will be ample. For Runners, one drake can easily look after eight ducks, but in order to be on the safe side, it is always advisable to have two drakes. I have known many cases of disappointment owing to a drake having proved sterile. If, therefore, it is decided to purchase, say, twenty-five ducks, these should be mated to five drakes, which will ensure good fertility. It is not advisable to purchase old drakes, unless it is certain they are not more than a season old. Old ones may be quite satisfactory in March and April, but a person anxious to work up a flock will wish to set eggs in February. Old drakes at that time of the year are not very ardent, unless the weather happens to be particularly mild, whereas the first season drakes show much more activity.

In May, in the case of the lighter breeds, two active drakes can easily keep as many as thirty ducks covered. A customer of mine, who hatched and reared several hundred ducks from eggs he had from me, tried the experiment of dividing up 150 ducks into three breeding pens. In pen A he put one drake, in pen B he put two drakes, and in pen C three drakes. A careful note was kept of the fertility percentage from each pen. A separate incubator was filled with eggs from each pen. The result was that eggs from pen A, with one drake mated to fifty ducks, showed 50 per cent fertility. Pen B, with two drakes and fifty ducks, gave 90 per cent fertility, whereas pen C, with three drakes and fifty ducks, only showed 70 per cent fertility.

Unfortunately, the presence of a lot of drakes among the ducks in the later part of spring, often leads to the ducks becoming badly mauled and often blinded. Taking several drakes out of the pen, may result in a number of ducks that had paired off with these drakes, refusing to be touched by the remaining drakes for some while, with the result that for a time a high percentage of unfertilized eggs are forthcoming.

The Best Choice to Make

Anyone commencing with twenty-five breeding ducks, can easily work up a flock of 300 layers by the end of the breeding season, if three or four incubators of 100 eggs capacity are kept going. To purchase eggs for hatching this number of ducks, would mean buying at least 1,000, because one must not reckon on hatching more than 50 per cent ducklings, and allow for 50 per cent drakes. The cost of these eggs would, in the case of Khaki Campbells or Runners, work out at about 30s. per hundred in peace-time, and that without replacement of clear eggs, unless an unusually high percentage were found to have been delivered. Anyhow, if we put down the cost of purchasing 1,000 fertile eggs at £15, we should not be far out. This is practically equivalent to the cost of the purchase price of twenty-five breeding ducks and five drakes.

Starting Costs

The cost of feeding the breeding stock, say from September to the following June, which is ten months, would be about £17. If the ducks have been properly fed during that time, they should have produced somewhere in the neighbourhood of 2,500 eggs, so that, after deducting the cost of the eggs you would otherwise have had to purchase at £15, and the sale of the balance 1,800 eggs at 1s. 5d. per dozen, roughly £10 10s., it will be seen that even with this moderate output of eggs it is as cheap to commence by buying breeding stock as it is to purchase eggs, and probably far more satisfactory. More satisfactory, because, as I have already mentioned, the eggs do not get the gentlest of handling by the railway people, and badly shaken eggs are, as a rule, bad hatchers. Further, in order to get the best results from artificial incubation of duck eggs, it is essential that they should not be more than six days old when put into the incubator, unless they have been carefully stored by wrapping them in greased paper and turning them every day. Please note that the eggs themselves must not be greased, as this is detrimental to their hatching powers. Stale ducks' eggs rarely hatch well—you may get a 50 per cent hatch, but probably not that. Eggs intended for hatching should never be thoroughly washed, but just carefully wiped over with a damp rag. Excessive washing removes the natural covering of grease, and causes excessive evaporation of the contents, resulting, in many cases, in a poor hatch.

If day-old ducklings are bought, it will probably cost about £35 to get your flock of 300 ducks, for £6 per hundred is what one would have to pay for ducklings bred from anything like good stock. These prices, of course, vary according to season and national circumstances. There is, however, a certain amount of risk in purchasing day-old ducklings, especially if the weather at the time of despatch is cold, and particularly if there is a biting north-east wind, which is often prevalent during February and March.

How to Avoid Chilling

If you are purchasing day-old ducklings from a breeder living at some distance, which means that they will have to be sent by rail, then, when you order the ducklings, it is essential to ask the sender to let you know by what train he is despatching them. Further, if he is a business-like person, he should give you the approximate time that they should arrive at your station.

So many ducklings are lost as a result of their being chilled en route, and they should, therefore, never be kept waiting about at the station for longer than is absolutely possible. My own practice was to send ducklings off overnight, despatching them by the latest possible train. If the ducklings are a few days old when despatched, by sending them at night one can send them properly fed and watered, with sufficient food in their crops to produce their own body heat for at least twelve hours. If despatched in the morning, they will not have sufficient food in them to produce the necessary heat.

When ordering the ducklings, be absolutely explicit in your directions as to where the ducklings are to go to, and write your name, address, and station address in capital letters, so that there will be no errors made. I have had so much experience in these matters, that I know only too well the difficulties that the vendor of ducklings is put to as a result of bad writing and the wrong railway address being given.

Help the Breeder

Remember, also, that when a breeder advertises his ducklings and guarantees " live delivery," he cannot be expected to do the latter if you do not collect the ducklings as soon as they arrive. After all, you have got to be honest as well as him, and it is not honest to write and tell him afterwards that some of the ducklings were dead on arrival, if you have delayed collecting them for several hours, as happens occasionally. Once a railway booking clerk very kindly wrote to me, to say that a customer to whom I had sent two dozen day-old ducklings had not collected them for twenty-four hours after they had arrived at the station. Both the booking clerk and I had sent postcards notifying the purchaser—I, of the time the ducklings could be expected, and he of their arrival. This customer later wrote to me an irate letter saying that only eleven ducklings were alive when he collected them from the station, and made no mention of the delay in fetching them, and he wanted me to replace the dead ones. I refused and referred him to the booking clerk at his station. I heard no more about the matter. But I know this to be no very unusual occurrence. Not only is it unfair to the vendor, but it is an extremely cruel thing to do to the ducklings.

It is also up to the vendor of the ducklings to send a postcard to the purchaser as to time of despatch, even if not asked to do so, and in the case of a big order, he should send a telegram at his own expense. The life of one duckling saved in transit will pay for the telegram. The information will also enable the purchaser to have an opportunity of seeing that everything is ready to receive them.

The prices mentioned in the foregoing chapter are all pre-war figures and to-day, in 1944, it must be reckoned that hatching eggs from reliable Khaki Campbell, Aylesbury or Runner stock birds will cost from 15s. to £1 per dozen : Day-old ducklings 40s. per dozen unsexed, and 65s. to 70s. per dozen sexed day-old ducklets ; unsexed week-old ducklings 50s. per dozen ; eight week old ducks 12s. each ; pullet ducks on point of lay, that is, when twenty weeks old, 25s. to 35s. each. A good drake will cost about 35s. to 60s.

Please note that pullet Khaki Campbells when four months old should have brownish or greenish bills, and not slatey coloured ones. If they have the latter, the chances are that the birds are nearer ten weeks old and have not yet completed their first juvenile moult.

MAKING A START

IT is only too easy to rush into anything, spend good money and then find you have got the wrong thing. In this short chapter I propose to put myself in the place of one who has suitable ground and wishes to commence in ducks.

How shall I make a start? Much depends on the season of the year. If in the autumn, I can start by the purchase of a trio, pen or even a small flock of correctly mated breeding stock. Depending on the breed, it should be possible to buy reliable breeding birds. It is better to acquire a few tip-top birds likely to breed perfect progeny than to rush in and purchase lots of inferior birds from all over the place.

A trio—a drake and two ducks, properly mated, first- or second-season birds, would last as foundation stock for a number of years and if used purely as breeders and not forced I would expect each season to rear from them a number of layers, the choicest of which would be breeders for the future.

Even if I bought stock, that would not prevent me from buying a few sittings of eggs at the spring of the year, toe-marking the resulting ducklings and maybe mating some of them up later on for the future.

When purchasing eggs you can generally take your choice of 12 eggs and " clears " replaced or 15 eggs and no replacements. Usually if 12 eggs are taken the seller will replace any " clear " eggs which are returned to him for examination, say, within 14 days.

A " clear " egg is one which has no signs of incubation and which when held before a light looks just like a fresh egg, except that the air space at the top will be larger. Such eggs will be replaced.

Eggs which have been fertilized by the male and then the

DUCKS

germ dies are not replaced. If you go to a reliable breeder and fancier, who values his name and reputation, my advice is take 15 eggs, one trouble in setting and in rearing the results. Most breeders will meet your request for less number of eggs or even mixed sittings from different breeds and varieties.

By going to a reliable breeder and buying birds, however, I would certainly have something to look at and to carry on with for the future; better still, they would be correctly mated and likely to breed good birds. If they did so I would

Ducks will find much natural food and will enjoy the shade afforded by the trees if allowed to run in an orchard. They will not damage bush fruit

leave the original pen for a number of years to work up into a large flock.

Alternatively, ducklings could be purchased, so there are a number of ways of making a start, depending on the season of the year and circumstances.

A drake and three, four or five ducks would give me more eggs to sit at the right season of the year. However, in the case of foundation stock I would increase the period of hatching, toe-marking and ringing all the progeny. The later

DUCKS

ones could be run on, say, as layers and later, when in the second season, they would prove reliable breeders.

With all these alternatives it is up to the would-be duck-keeper to use his own ideas and judgment as to the best method of starting. Make out a plan and go right out to carry it through to success. Just one other thing—try to make up your mind once and for all on the breed you wish to keep, do not keep chopping and changing, it costs good money and wastes time. Get on to a good breed which you like and which is suitable for your purpose and surroundings and stick to it. Get to know all you can about the breed and try to improve it for yourself and for the good of the breed.

As later chapters will show, there are other egg-laying ducks than Khaki Campbells and Indian Runners, while the Aylesbury need not necessarily be regarded as the only table type of duck.

I like the beginner who sets himself a programme and sticks to it. He is, in fact, the one type of person likely to succeed. Failures are usually found among those who continually chop and change and never appear to be satisfied with what they are doing.

First-Class Breeding Stock.

The above shows the necessity of first-class breeding stock to start with. I do not mean fancy stock at all, as many of the points of excellence claimed by the American standard militate directly against the market value of the birds. A few years ago several men came here to buy Pekin ducks for breeding stock. On looking at the birds and getting the price, one man said: "Those are the best birds I ever saw. I want thirty of the best birds you have." Another said: "They are fine birds, but I cannot afford to pay two dollars for a duck; have you no cheaper birds?" "Yes, I have some later birds—culls from which the rest have been selected. They are not as large as these. My late birds never attain the size of the earlier-hatched ones, and they will not lay quite as early. You can have your choice of these at one dollar each, which is about their market value."

He took those birds, and I consider when he made that choice that he threw away more than $100 of his first season's work alone, for, with a fair share of success he might easily expect to raise 100 young birds from each of his breeding ducks, and as the birds he chose were at least one-third lighter than those he rejected, their progeny would not be as heavy at a marketable age by at least one pound per bird. The excess in cost to him, had he bought the better birds, would have been but one cent on each of the young birds he raised. He lost, on making the choice he did, more than twenty cents on each bird, and this is not all; those birds will be small for generations to come. He never can get them up to the standard of the others. They will go upon the market as small birds, and as such, command at least two cents per pound less than the larger ones; in fact, his losses in this transaction will represent a large share of the profits.